The Darwin Archipelago

ALSO BY STEVE JONES

The Language of the Genes

In the Blood

Almost Like a Whale

Y: *The Descent of Men*

The Single Helix

Coral: A Pessimist in Paradise

The

Archipelago

The Naturalist's Career Beyond *Origin of Species*

STEVE JONES

Yale UNIVERSITY PRESS NEW HAVEN & LONDON

Published with assistance from the Louis Stern Memorial Fund.

An earlier edition of this book was published in Great Britain in 2009 by
Little, Brown as *Darwin's Island: The Galapagos in the Garden of England.*

Yale University Press books may be purchased in quantity for educational, business, or
promotional use. For information, please e-mail sales.press@yale.edu (U.S. office)
or sales@yaleup.co.uk (U.K. office).

Set in Electra Roman type by Tseng Information Systems, Inc.
Printed in the United States of America by Sheridan Books, Ann Arbor, Michigan.

Library of Congress Cataloging-in-Publication Data
Jones, Steve, 1944–
[Darwin's island]
The Darwin archipelago : the naturalist's career beyond Origin of species / Steve Jones.
p. cm.
Includes bibliographical references and index.
Originally published: Darwin's island. London : Little, Brown, 2009.
ISBN 978-0-300-15540-2 (hardcover : alk. paper)
1. Darwin, Charles, 1809–1882. 2. Naturalists—Great Britain—Biography.
3. Evolution (Biology). 4. Natural selection. I. Title.
QH31.D2J76 2011
570.92—dc22
[B]
2010038593

A catalogue record for this book is available from the British Library.

This paper meets the requirements of ANSI/NISO Z39.48–1992 (Permanence of Paper).

10 9 8 7 6 5 4 3 2 1

To R. C. Simpson, who first taught me biology

Contents

Preface ix

CHAPTER 1. The Queen's Orang-Utan 1

CHAPTER 2. The Green Tyrannosaurs 26

CHAPTER 3. Shock and Awe 46

CHAPTER 4. The Triumph of the Well Bred 69

CHAPTER 5. The Domestic Ape 90

CHAPTER 6. The Thinking Plant 113

CHAPTER 7. A Perfect Fowl 130

CHAPTER 8. Where the Bee Sniffs 150

CHAPTER 9. The Worms Crawl In 166

Envoi: Darwin's Island 187

Annotated Bibliography 205

Index 219

Preface

CHARLES DARWIN, AS MOST PEOPLE KNOW, SAW THE FINCHES OF the Galapagos in the years he spent in the archipelago while employed as official naturalist on HMS *Beagle*. Each island had its own species, and Darwin soon worked out that they shared descent from a common ancestor; that they were a product of evolution. On his return to England he at once published his theory in his book *Origin of the Species*, in which he went on to prove that humans had descended from chimpanzees. Nature, red in tooth and claw, had used the survival of the fittest to weed out the imperfect and, with *Homo sapiens* at the top of the evolutionary tree, had achieved her desired end. Racked by guilt at replacing the doctrines of the Church with a joyless vision of man as a shaven primate in an amoral universe, Darwin retired into obscurity. He repented his blasphemy on his deathbed and was buried as a venerable and almost forgotten savant who—like so many famous scientists—had done his most important work as a young man.

All this is an entire parody of the truth. Darwin was not a hired biologist but paid for his own trip as gentleman-companion to the *Beagle*'s captain. He spent just five weeks of the five-year voyage in the Galapagos and set foot on only four of its dozen or so islands. He had little interest in his collection of finches and lumped their corpses together as a jumbled mass without even noting from which island they came. Two decades passed before the publication of *The Origin of Species* (in which the word "evolution" does not appear), and during that period its author wrote several substantial books. The phrase "the survival of the fittest" is not his but was invented by the philosopher Herbert Spencer to summarise the notion of natural selection, the central element of evolutionary theory. The bloody fangs and fingernails of Mother Nature were thought up by Tennyson a decade earlier and not as a philosophy of life but in memory of the death of a friend. Evolution has no end in view and men do not descend from chimps, although the two share a common ancestor (an idea not explored by Darwin for a dozen years after *The Origin*). The Church soon accommodated his ideas, which, as its clerics realised, have no relevance to religion. His deathbed conversion is a simple falsehood (even if the great naturalist was buried in Westminster Abbey, where he still lies, trampled by tourists).

The most widespread error is to assume that the *Beagle* voyage marked the end of Charles Darwin's scientific career. In the four decades that remained to him after he came back from the wilds in 1836, Captain Fitzroy's gentleman-companion worked as hard as or harder than he had as a young man. He soon purchased Down House, south of London. At first he thought the place dull and unattractive, but before long it was transformed, with the help of his considerable fortune, into a grand but comfortable mansion. Its owner settled in the land of his birth and never again left: uxorious, paternal, and reluctant to abandon his extensive garden except on forays to test his theories and, now and again, to search for better health. Great Britain was the first and last of his forty islands and he studied its natural history in far more detail than that of anywhere else. For him, Kent was as much a place of discovery as had been the jungles of the Amazon or the stark cinders of the Galapagos. The British Isles were where Charles Darwin built his reputation.

This book is about the disregarded Darwin and about his years of work on the plants, animals, and people that make their home in the land of his birth. *The Origin of Species* is without doubt the most famous book in science. It celebrated its hundred-and-fiftieth anniversary in 2009 (which also marked the author's bicentenary), but to remember Darwin for that magnum opus alone would be as foolish as to celebrate Shakespeare only as the author of *Hamlet*. His lifelong labours—six million words in nineteen published works, hundreds of scientific papers, and fourteen thousand letters—generated an archipelago of information, a set of connected observations that together form a harmonious whole. Biology emerged from that gargantuan effort as a unitary subject, linked by the great idea of common ancestry, of evolution. The volumes written in Down House made sense of a whole new science and enabled its students to navigate what had been an uncharted labyrinth of shoals, reefs, and remote islets of apparently unrelated facts.

The Origin is a work of reportage as much as of research. Most of Darwin's other books are, in contrast, based on his own observations and experiments. They explore, with trademark enthusiasm, what seem at first sight to be almost unrelated aspects of the natural world. Charles Darwin's domestic works, as they might be called, are (in order of publication and with titles somewhat truncated): *Barnacles* (in four volumes), *Orchids and Insects*, *Variation under Domestication*, *The Descent of Man*, *Expression of the Emotions*, *Insectivorous Plants*, *Climbing Plants*, *Cross and Self-Fertilisation*, *Forms of Flowers*, *Movement in Plants*, and *Formation of Vegetable Mould by Earthworms*. Much of his oeuvre was aimed at a wide audience and is set out in good, plain Victorian prose. He wrote to Thomas Henry Huxley that "I sometimes think that general and popular Treatises are almost as important for the progress of science as original work." Charles Darwin was the era's greatest popular science writer—and his publisher realised as much for he gave *The Origin* equal billing with Samuel Smiles's quintessentially Victorian work, *Self-Help*, which appeared on the same day.

Here I attempt to bring his writings up to date for the modern age and to place the world's greatest biologist firmly in the context of his native land. His literary canon makes sense only when considered as a whole. At first sight its subjects seem disconnected—earthworms, inbreeding, barnacles, plant hormones, domestication, insect-eating plants, the expressions of joy or despair in dogs, apes, and men—but in truth all share a theme: the power of small means, given time, to produce gigantic ends. Fond family man as he was, he saw no gulf between the powers that had made his wife and children and those at work throughout the living world. His concerns about the risks of marrying his cousin were tested with experiments on flowers, and his interest in the emotions of animals extended to comparing the expressions of his infant son to those of dogs and apes. All had emerged through the action of the same biological process, through evolution or "descent with modification," a notion that outraged some of his contemporaries (and leaves many people uncomfortable today).

Biology has plenty of heroes, but Charles Darwin is unique, for he was a pioneer in so many of its branches. He became a better scientist as he grew older for he began to test ideas with experiments, many far ahead of their time, rather than collating the results of others, brilliant as the synthesis might be. A good portion of the educated public has heard of *The Origin* and *The Voyage of the Beagle*, but his other works are almost unknown. Biologists tend to be familiar with at least some of them, for each volume is a milestone in their profession. The *Earthworms* epic founded modern soil science, *Emotions* saw the dawn of comparative psychology, and *Cross and Self-Fertilization* and *Forms of Flowers* were early attempts to understand the origin of sex. The experiments described in *Movement in Plants* gave the first clue to the existence of hormones (although the word had not been invented and their discovery in animals had to wait thirty years). Darwin also wrote on carnivorous plants, on the links between insects and orchids, and on the origin of crops and domestic animals (and there he grappled with the nature of heredity and almost got it right, with talk of crosses between round and wrinkled peas). His four books on barnacles showed that juvenile forms reveal more about relatedness than do adults, and that bodies as complicated as our own are built on a simple plan. For barnacles, as for everything else, natural selection made organs of great sophistication not by design but by tinkering with whatever raw material was available.

The Descent of Man, and Selection in Relation to Sex, to give its full title, stands rather apart from the rest. It appeared in 1871 and was both the first real treatment of human evolution and an introduction to the evolutionary importance of sexual conflict. It sets out the entire evolutionary argument with reference to a single group of creatures: man and his relatives. Like *The Origin of Species*, it is in the main a compilation of the results of others. Even so, it fits well into what might be called the Down House School, and I use it here as an introduction to the world of modern biology. The study of our past has been transformed since it was written. If the author of *The Origin* were to rewrite that book today, he would turn for many of his examples not to pigeons and tortoises, nor to worms and barnacles, but to his fellow men. *The Origin*'s only mention of *Homo sapiens*, the tentative claim that "Light will be thrown on the origin of man and his history," has been wonderfully upheld. It shows how the truths glimpsed by Darwin now unite every part of the study of life, our own included.

Here I attempt to update all these topics. I append an envoi with a look at the biological world of the twenty-first century compared to its state in 1859. I avoid as much as possible any account of the relevance of Darwinism to the human predicament and the empty arguments about its interactions with religion. Most scientists have no interest in the struggle to separate science from theology (although there are exceptions: Thomas Henry Huxley felt that "extinguished theologians lie about the cradle of every science as the strangled snakes besides that of Hercules"). Today's biology emphasizes how little relevance the subject has to the issues so often and so tediously discussed by nonbiologists. As Darwin put it in *The Descent of Man*, "We are not here concerned with hopes or fears, only with the truth as far as our reason allows us to discover it." Science can do that, and no more.

My eminent predecessor at University College London, the Nobel laureate Peter Medawar, in an acerbic comment on the relative merits of students of science and the arts, said of Watson and Crick (of double helix fame), "Not only were they clever, they had something to be clever about." Not only did Charles Darwin travel, he had something to travel for. The joy of the *Beagle* voyage was that it had a point. For a real adventurer, to travel hopefully is not enough: some end must be in view. As he says in the last pages of his account of the journey: "If a person asked my advice before undertaking a long voyage, my answer would depend upon his possessing a decided taste for some branch of knowledge, which could by this means be advanced."

Darwin's voyages, from the Galapagos to West Wales, play an important part in *The Origin*'s "long argument," as they did in its author's life. The *Beagle* crossed thirty thousand miles of ocean, but his British journeys covered almost as much distance. His work was always tied to where he found himself, whether in a rain forest or a suburb. His writings comprise a kind of Grand Tour of the British Isles. His very first memory, as recounted in his *Autobiography*, was of visiting Abergele for the sea-bathing at the age of four. Six years later he was back on the Welsh coast at Towyn, where he noted for the first time some "curious insects" (perhaps the black and red Burnet Moth) and, unlike the many naturalists of those times who filled cabinets with butterflies or shells to make a biological stamp collection, he wondered, even as a child, why they were found in one place and not another.

His juvenile interest in the insects of England and Wales grew into a lifelong exploration of the biology and geology of his native island. He published his first scientific paper (on the eggs of an animal found in the Firth of Forth) during the twelve chilly months he spent in Edinburgh. After a brief visit to Dublin, the young enthusiast then moved to Cambridge, where he spent days wading through bogs and fens in search of specimens. Just before setting off on the *Beagle*, he travelled for three weeks across North Wales from Shrewsbury to Conwy and Barmouth with the geologist Adam Sedgwick, who taught him the elements of mapping that proved so useful on the voyage. On his return he set off again to Scotland, where, in his first major scientific paper, he made a frightful error in interpreting a series of parallel shelves or "roads" in Glen Roy as wave-cut beaches rather than as the shores of drained glacial lakes (as he wrote much later, "I am ashamed of it"). Later in life he crisscrossed the countryside to pursue his researches or to take his family on holiday; to Wales, the Isle of Wight (where he met Alfred, Lord Tennyson), to Torquay, to the Lake District (a meeting there with Ruskin), to Stonehenge, to the heathlands of England, and to a variety of grand mansions across the nation. In his forty years at Down House, Charles Darwin spent two thousand nights away from home—the equivalent of a day for each week he lived there. A few of his trips lasted a month or more.

Some of his travels were in search of science, but many were a quest for health. He visited spas in Great Malvern, Guildford, and Ilkley (where he received the first copy of *The Origin*). His later years were marked by a series of bizarre attempts to remedy his feeble state (even if he did write that illness, "though it has annihilated several years of my life, has saved me from the distractions of society and amusement").

The main symptom was vomiting. So severe were the attacks that he declined some invitations to stay in friends' houses on the grounds that "my retching is apt to be extremely loud." He tried Condy's Ozonised Fluid, "enormous quantities of chalk, magnesia & carb of ammonia," and rubber water bags worn next to the spine. Nothing worked. The author of *The Origin* was one of the victims of the Victorian "Demon of Dyspepsia" (an unhappy throng that included Thomas Carlyle, George Eliot, Charles Dickens, Florence Nightingale, and the evolutionists T. H. Huxley, Alfred Russel Wallace, and Herbert Spencer, together with his own brother Erasmus). Their troubles funded several pharmaceutical fortunes, including that of Henry Wellcome, which later helped pay for that Darwinian triumph, the sequence of human DNA. The cause of his illness is not known: a conflict between Christian belief and rationalism, a parasite picked up in Brazil, an inherited sensitivity to raw milk, or even, some say, the obsessive swallowing of air. Dyspepsia's symptoms seem closest to those of a stomach ulcer, whose nausea, depression, and lassitude are, we know today, caused by a bacterium. The bug that swept through Victoria's intellectuals might now be cured with a simple pill.

Later in life the aging paterfamilias spent longer and longer periods without leaving his own house. He fed his family fifty-three distinct varieties of gooseberries and three of cabbage. In his garden he carried out many experiments, helped by William Brooke, his "gloomy gardener" (who was seen to laugh only once, when a boomerang broke a cucumber frame). The great naturalist's tale ends, in the tradition of the classics, with the hero's death and his desire to join his beloved earthworms in the "sweetest place on Earth," the village churchyard at Downe—a wish frustrated by fame and the Abbey.

The Darwin Archipelago retraces some of Darwin's steps. It will, I hope, help bring his less-well-known work into the twenty-first century. Several people have helped in the preparation of this book. David Leibel, Michael Morgan, and Anna Trench made helpful comments on parts of it. I thank them for their help.

Three of my earlier volumes—on coral reefs, on the nature of maleness, and on the theory of evolution itself—pay homage to the founder of the science of life by attempting to update his ideas for the modern age. There could be no better way to honour the most famous of all biologists than to give his less celebrated writings the exposure they deserve. For Charles Darwin, the five *Beagle* years that became part of Britain's intellectual legacy led to four decades of intense labour within the confines of his native land. In that modest group of islands he underwent a second great voyage: not of the body but of the mind. This book traces his journey from its beginning to its end.

The Darwin Archipelago

CHAPTER 1

The Queen's Orang-Utan

IN 1842 QUEEN VICTORIA WENT TO LONDON ZOO. SHE WAS LESS THAN amused: "The Orang Outang is too wonderful . . . he is frightfully, and painfully, and disagreeably human." The animal was not a male but a female called Jenny, and Charles Darwin had, some years earlier, visited its mother. He too spotted the resemblance between the apes on either side of the bars. The young biologist scribbled in his notebook: "Man in his arrogance thinks himself a great work. More humble and I believe true to consider him created from animals." Seventeen years after Victoria's visit, in 1859, he published the theory that established the Queen's and his own kinship to Jenny, to every inmate of the Zoo, and to all the inhabitants of Earth.

The Origin of Species led to an uproar among the Empress of India's subjects. Her Chancellor, Benjamin Disraeli, asked: "Is man an ape or an angel? My Lord, I am on the side of the angels. I repudiate with indignation and abhorrence these new fangled theories." Many of his fellow citizens agreed. Even so, in time, and with some reluctance, the notion that every Briton, high or low, shared descent with the rest of the world was accepted. A quarter of a century later, W. S. Gilbert penned the deathless line "Darwinian man, though well behaved, at best is only a monkey shaved," and the idea of *Homo sapiens* as a depilated ape became part of popular culture. Victoria herself congratulated one of her daughters for turning to *The Origin*: "How many interesting, difficult books you read. It would and will please beloved Papa."

As the Queen had noticed, the physical similarity of men to apes is clear. In 1859 Charles Darwin came up with the reason why. A certain caution was needed in promoting the idea that what had made animals had also produced men and women, and he waited for twelve years before expanding on the subject. *The Descent of Man* describes how—and why—*Homo sapiens* shares its nature with other primates. It uses our own species as an exemplar of evolution.

To the great biologist, man obeyed the same evolutionary rules as all his kin—in the book's final words, he reveals "the indelible stamp of his lowly origin." In moral terms *Homo sapiens* was something more: "Of all the differences between man and the lower animals, the moral sense or conscience is by far the most important. This sense . . . is summed up in that short but imperious word 'ought,' so full of high significance. It is the most noble of all the attributes of man, leading him without a moment's hesitation to risk his life for that of a fellow-creature." No ape understands the meaning of "ought," a word pregnant with notions quite alien to all species except one: but despite that essential and uniquely human attribute, every ape, like every other creature, is the product of the same biological mechanism.

The logic of evolution is simple. There exists, within all creatures, variation, which is passed from one generation to the next. More individuals are born than can live or breed. As a result, there develops a struggle to stay alive and to find a mate. In that battle, those who bear certain variants are more successful than others less well endowed. Such inherited differences in the ability to pass on genes—natural selection, as Darwin called it—mean that certain forms become more common as the generations succeed. In time, as new and advantageous versions accumulate, a lineage may change so much that it can no longer exchange genes with others that were once its kin. A new species is born.

Natural selection is a factory that makes almost impossible things. It generates what seems like design with no need for a designer. Evolution builds complicated organs like eyes, ears, or elbows by piecing together favoured variants as they arise and, almost as an afterthought, produces new forms of life. Its tale as told in *The Origin of Species* turns on the efforts of farmers as they develop new breeds from old, on changes in wild plants and animals exposed to the rigours of nature and the demands of the opposite sex, on the tendency of unique forms to evolve in isolated places, and on the silent words of the fossils that tell of a planet as it was before evolution moved on. The book's pages also speak of the embryo as a key to the past and of how structures no longer of value and others that appear unreasonably perfect are each testimony of the power of natural selection. The geography of life—on islands, continents, and mountains—is evidence of the common descent of mushrooms, mice, and men. Most of all, the diversity of the living world can be arranged into a series of groups arranged within groups, of ever-decreasing affinity, proof that their members split apart from each other longer and longer ago.

The Descent of Man uses that logic to disentangle the history of a single species. Unique as it might think itself, *Homo sapiens* is an animal like all others. The book's famous last sentence reads, in full: "I have given the evidence to the best of my ability: and we must acknowledge, as it seems to me, that man with all his noble qualities, with sympathy which feels for the most debased, with benevolence which extends not only to other men but to the humblest living creature, with his god-like intellect which has penetrated into the movements and constitution of the solar system—with all these exalted powers—Man still bears in his bodily frame the indelible stamp of his lowly origin."

In 1871—and even in 1971—the evidence for that final and provocative statement was weak indeed. Now, everything has changed. The entire evolutionary case can be made in terms of ourselves and our relatives; of apes and monkeys, chimps and gorillas, and men and women. The new biology means that we know more about the human past than about that of any other species. Evolution is best viewed through our own eyes, not just because we are interested in where we came from but because advances in science mean that *Homo sapiens* has become the embodiment of every evolutionary idea. The theory itself has not altered much in the century and a half since it was proposed. The technology used to study it has.

Complicated as they are, some of the tools used today to examine the past would have been familiar in their nature, if not in their details, to biologists of the nineteenth century. Charles Darwin was, among his many talents, a great anatomist. He used changes in the physical structure of pigeons, pigs, and people as evidence for his theory. The first chapter of *The Descent of Man* is a somewhat ponderous comparison of the bones and bodies of men and apes. Dissection, once at the centre of biology (and biologists of a certain age still flinch at the smell of formalin), not long ago seemed antiquated, but now looks very modern. Molecular biology is no more than comparative anatomy plus a mountain of cash. Its chemical scalpels cut up creatures into thousands of millions of letters of DNA code. Those who wield them have shown beyond all doubt the truth behind Queen Victoria's fear that Jenny the orang-utan was proof of the common ancestry of humans, orang-utans, and more.

The human genome project—the scheme to read off our own DNA sequence—set the seal on an enterprise that began in the sixteenth century, when Vesalius opened a human heart and discovered that it had four chambers rather than the three insisted on by Galen. Its completion was announced in 2000 and again in 2003, 2006, and 2008 (and some parts of the double helix still remain unread). A science that in its infancy was a mere description of bones and muscles became an adolescent when *The Origin of Species* showed how shared structure was evidence of common descent. It has at last matured, and the anatomy of DNA has become the key to the history of life.

In a glass-fronted cabinet at University College London resides the stuffed body of the eighteenth-century philosopher Jeremy Bentham, the "greatest good for the greatest number" man. His Auto-Icon, as he called it, was an attempt at a memorial that would cost less than the showy shrines then fashionable. Bentham was convinced that his idea would catch on. Two centuries later, it did. James Watson—one of the duo who unwound the double helix—was presented with his own auto-icon, a compact disc bearing his entire DNA sequence, which he can, if he wishes, display for public edification in a small plastic case.

Watson's essence is coded into a tangled mass of intricate chemistry. The egg that made him contained two metres of DNA, and each of the billions of cells that descend from it as his body grows, ages, and dies has a copy. Every one of those molecular sentences is around 3.2-billion letters long, the letters consisting of the four bases of the familiar genetic code. Twenty years ago, when the scheme to read the whole lot was proposed, it took months to decipher the number of letters found in this paragraph. The molecule was sliced into random bits, each was read from end to end, and the whole genome was stitched together with a search for places where the fragments overlapped. Such methods have become antique. Today's machines pick up flashes of light from molecules tagged with fluorescent dyes, each base with its own colour. It takes no more than a few hours to read off a piece as long as this entire book (which itself contains less than one part in several thousand of the whole content of the human genome). Now, enthusiasts speak of machines that will read off our entire genome in an hour.

The first human sequence cost a billion dollars, and Watson's version was auctioned off for a million. In 2009 the Knome Corporation offered on eBay to read off the DNA of anyone willing to pay the starting price of $68,000. In fact, the whole lot can already be done for a twentieth of that sum. Within five years the price will drop even further. IBM hints that it will soon be able to do the job for $100, and it will then become possible to decipher the DNA of any creature at nominal cost.

The raw material of evolution in its physical structure as a double helix is simple and elegant, but its biology is not. It shows that life did not emerge from engineering but from expedience. In its details, DNA is, frankly, a mess, for natural selection can build only upon what it already has. The Darwinian machine has no strategy and can never look forward. Its tactics are based on the moment, and the genome, like the creatures it codes for, is the product of a series of short term fixes. James Watson's molecule is marked by redundancy, decay, and the scars of battles long gone. The chaos within transcends even that of the bodies it builds. Genes—like cells, guts, and brains—work, but only just.

Human DNA contains great stretches that appear to be useless, and numerous sections that mirror each other. Repetition is everywhere: of particular genes into families that carry out related tasks and of multiplied lengths of apparently redundant material. The remnants of viruses make up almost half the total, and much of the rest is littered by the decayed hulks of ancient and once functional structures. All but one part in fifty were once dismissed as biological garbage.

Genes have become blurred and ambiguous as we learn more. They are far fewer in number than expected when the genome project was proposed—just over twenty thousand rather than the hundreds of thousands once believed to be essential. Some overlap each other or say different things when read in opposite directions. Many contain inserted sequences that seem to have no function (although some of the supposed junk does a useful job and other sections cause disease should they wake up and shift position). Plenty of questions remain. How important is the segment that codes for proteins, compared with the on and off switches, the accelerators and brakes, and the rest of the control machinery? We do not know.

Even the size of the package makes little sense. A chicken has less DNA than a Nobel laureate, although half its genes are identical (or almost so) to our own—evidence, given that we last shared an ancestor three hundred million years ago, of how conservative evolution can be. A tiny plant called *Arabidopsis*, a relative of the Brussels sprout, has more genes than either. This may say more about how hard it is to define a gene than about the talents of sprout versus Nobelist.

Eight decades passed between Vesalius's dissection of the heart and

William Harvey's discovery of the circulation of the blood. The genome is now in that transitional period. DNA's nuts and bolts (and even some of its bells and whistles) have been dismantled, but most of those who work on it still study structure without much insight into function. Harvey saw the heart as a mere pump and understood nothing of its exquisite system of control. Genes are much the same. Each is linked into a network with others and responds to messages from within and outside the cell. The path from instruction manual to product is a labyrinth rather than a straight line. The proteins that pour from the cell's biological factories are not simple blocks that slot together but are entities that are folded, spliced, cut, or fused into new mixtures in a way that depends almost as much on local conditions as upon their own structure. Diseases as different as diabetes and prostate cancer may arise from damage in the same segment of DNA, while others, such as breast cancer, emerge from errors in several different sections. Most of the double helix is switched off most of the time, African genes are on average more active than those of Europeans, and life is beginning to look more complicated than molecular biologists once hoped.

Evolutionists are not in the least surprised. They were baffled at some of the decisions made by those who designed the genome project. Like Vesalius, James Watson and his colleagues took a Platonic view of existence. Every heart and every human was built on the same plan, and to understand one was to understand all. The early DNA sequencers out-Platoed Plato in assuming not just that the essence of humankind could be expressed in a single DNA sequence but that this sequence could be stitched together out of bits of double helix, taken from random donors across the globe, into a kind of genomic Everyman.

That was a mistake. The Platonic approach ignores the vital truth that natural selection depends on difference. To understand our past, biology needs not one genome but many. We can see how, when, and where evolution has been at work only by mapping variation from person to person, from place to place, and from species to species. So central is diversity to the idea of descent with modification that the first two chapters of *The Origin* are devoted to its extent in the bodies and habits of plants and animals. Now genetics can tell the tale in the language of DNA.

James Watson's auto-icon disclosed no more than half his secrets, for it contained only one of the two versions of the double helix present in each cell. His rival in the race to decipher the genome, the biologist and businessman Craig Venter, was less reticent. He read off both his copies, that received from his father and that from his mother. Venter was happy to reveal its contents: his father died young of a heart attack, and he has himself inherited a variant that predisposes to that disease. He has also inherited gene variants supposed to increase the wish to seek novelty and to be active in the evening rather than early in the day, as well as a version that codes for wet rather than dry ear wax.

Whatever Venter's intimate chemistry says about his bedtime, his personality, or his auditory canal, it has a message for us all, for it gave the first hint of the true level of human diversity. Both his parents are white Europeans (and hence represent just a small sample of mankind), but their DNA sequences are distinct from each other at around one site in two hundred along the entire chain—which adds up to tens of millions of differences between them.

On a global scale, hundreds of millions of sites in the inherited alphabet vary from person to person. The "Thousand Genomes Project," now almost complete (and involving twice that number of human subjects), has set out to find out how many there are. Unlike its predecessors it will search out rare variants, those carried by fewer than one person in a hundred—and given the advances of technology, the project may cost no more than a few million dollars. It started with a detailed survey of individual Nigerians, Japanese, Chinese, and Europeans and is expanding its reach to include a less precise search for diversity in a wide range of other populations. Already it has shown that variation exists in vast abundance. Each chromosome, in every population, contains millions of single-letter changes. The variable sites are so closely packed together that, over short lengths of the double helix, they rarely separate when the molecule is cut, spliced, and reordered as sperm or egg is formed. Such long blocks represent sets of chemical letters that travel down the generations together. Rather like surnames (which are much the same thing on an enlarged scale), they are excellent clues to relatedness.

As well as the single-letter changes, the double helix is marked by duplications of certain pieces and deletions of others. The order of its letters may be reversed, and great stretches can hop to a new place. Some genes are arranged in families: groups of similar structures that descend from a common ancestor and have taken up a series of related jobs. The biggest—eight hundred strong—builds the senses of taste and smell. Its members vary in number from person to person, and some lucky individuals have fifty more copies of a certain scent receptor than do others.

Most changes involve fewer than ten letters, but some are a million basepairs from end to end. A few people may, because of the gains and losses, have millions more DNA bases and thousands more genes than do others (and the potential variation in dose represents more than the length of the largest human chromosome). Even so, some of the repeated segments have just the same structure in humans as in the coelacanth, from which we diverged four hundred million years ago.

DNA is a labile and uncertain molecule. Many mistakes are made when a multiplied sequence is copied as egg or sperm are formed, to produce longer or shorter versions than the original. Some sections move or multiply at a rate of one in a hundred each generation, rather than the one in a million once thought to be typical. Age changes us, and the double helix is re-ordered, duplicated, and deleted as the years go by. As a result, the offspring of elderly parents inherit more mutations than do those of young.

Biologists have long known that, with the exception of identical twins, everyone in the world is genetically distinct from everyone else and from all those who have ever lived or ever will. That claim is too modest. In fact, every sperm and every egg ever made by all the billions of men and women who have walked the Earth since our species evolved is unique. Such variety links individuals, families, and peoples into a shared network of descent. It shows how man is related to chimpanzees, gorillas, orangs, and macaques, and for that matter to plants and bacteria. Evolution, like astronomy, has always looked at the past through the eyes of the present, but its new technology—like the stargazers' giant telescopes—means that it can now see farther and deeper into the universe of life than ever before.

Even so, biology is not physics. The images that flood from the gene-sequencing machines are often blurred. Many statements about ancestry are filled with unproven (and often unstated) assumptions about the rate of change in DNA, the size of ancient populations, and the effect—or, too often, supposed lack of effect—of mutations on the well-being of those who bear them. The information in the genome is almost limitless, but its language is often hard to understand.

Fortunately, we have better witnesses to the truths of years gone by. Like the remnants of stellar rocks that sometimes strike the Earth, they are silent, shattered, and few in number, but at least they give direct evidence of how the past unfolded. Darwin was well aware of the importance of fossils to his case. One page in six of *The Origin* are devoted to the record of the rocks, its imperfections, and the central role it plays as proof of the fact of change. In 1871, no human fossils (with the exception of a skull now known to come from a Neanderthal) had been recognised. Now the primate record is far more complete than it was even a few decades ago. The tale it tells is fragmented and ambiguous, but what we know fits quite well with that revealed by the double helix.

In the Miocene era—from around twenty-three million to five million years ago—the Earth was a true Planet of the Apes. Primates were all over the place, with a hundred or more distinct species in Africa, Asia, and Europe. They lived in woodlands, plains, forests, and swamps. Some were no bigger than a cat while others grew larger than a gorilla. For much of the time their capital was in Europe, and many of our predecessors laid their bones there. Then the apes moved on, to set up shop in Africa. A ten-million-year-old fossil from Kenya may be the common ancestor of men, chimps, and gorillas. If so, it confirms Darwin's speculation that it was more probable "that our early progenitors lived on the African continent than elsewhere" (although he did not know, of course, that continents had broken up and drifted across the world, and that Africa itself did not exist in the earliest days of human evolution).

Almost all the players in that ancient drama have left the stage. The day of the apes gave way to a long twilight—now fast turning into night. The sun began to set on their family well before humans appeared, but, once they did, their nemesis was assured.

In 2009, the details of a crucial human ancestor were revealed. *Ardipithecus*—whose remains had been found fifteen years earlier in the Afar region of Ethiopia, a site famous for early primates—lived around four and a half million years ago, just a couple of million years after the split from the ancestor we share with chimpanzees. It already looked remarkably human. Its brain was still no larger than that of a modern female chimpanzee, but its feet suggest that already the creature could walk upright. The difference in size of the canine teeth between males and females is less than in today's chimpanzees, a hint that perhaps our sexual behaviour had already become less fraught than that of our closest living relative.

Lucy, the famous fossil of *Australopithecus afarensis*—a creature even more human in appearance, lightly built and only a metre or so tall, with relatively long legs and small teeth—was found only fifty miles away. She belonged to a group who lived around three million years ago. Others among her kin left footprints in Tanzanian volcanic ash as proof that they were walking upright without difficulty at a time their brains were only a third the size of our own. The males were still considerably larger than the females. *Homo habilis*—"handy man"—lived in South and East Africa for about a million years, starting two and a half million years ago. It had long arms, brow ridges, and a larger brain than Lucy's, and was reasonably good at making tools. Similar creatures made a home in Africa and others in Georgia. *Homo erectus*, the upright human, the next fossil claimed as a direct (or almost direct) human ancestor, emerged around 1.8 million years ago and may have split into two species in its territories in Africa and Asia. Some individuals had brains as large as our own and lived as far north as southern France. A slightly younger European arrived around 1.2 million years ago and left a few of his bones in the caves of the Sierra de Atapuerca in northern Spain. That ancient European, whose kin spread as far north as Norfolk in eastern England, was christened *Homo antecessor.* It may be the common ancestor of ourselves and the Neanderthals. A later European from around half a million years ago, Heidelberg Man, was probably a Neanderthal antecedent with no human descendants. He too first appeared in Africa.

Many (perhaps too many) more supposed members of our lineage have been named, and the human pedigree is beginning to look more like a bush than a tree, making real patterns of descent hard to trace. For much of history, our ancestors shared their world with species that were much more closely related to them than the chimpanzee is to us. Those days have gone and most of those members of our ancient family have left no issue today.

The Neanderthals, once our immediate kin, filled Europe and the Middle East from around two hundred thousand to about thirty thousand years ago. They had bigger skulls—and perhaps bigger brains—than modern humans (although they were also beefier in general). Their DNA looked remarkably like our own. The Neanderthals trapped animals in pits, and may have been cannibals (those who would prefer to consider them a less grisly relative suggest that the carved bones found in their caves may instead be remnants of a ritual burial). They lived in small groups in an icy Europe for far longer than our own species has existed; and then they disappeared. Like many other apes, they did so quickly. Perhaps a cold snap defeated them, for a remnant hung on in the warmth of southern Spain until well after the moderns arrived. The latter had better clothes, which allowed the tropical ape that they were—and we are—to survive in a climate that killed off an animal more used to bad weather but less well clad. Perhaps *Homo sapiens* murdered the Neanderthals or starved them out, but we do not know. Sex may even have been on the agenda, with evidence from Neanderthal DNA that a few of their genes may live on in modern humans. The double helix suggests that their last common ancestor with us lived in Africa as much as six hundred thousand years ago—long before *Homo sapiens* emerged.

Soon after the disappearance of his cousins, that species began to spread across the world. Modern humans filled the habitable globe only a thousand or so years ago, when men and women reached New Zealand and Hawaii. Their ancient journeys can still be read in DNA. It reveals a clear split between Africa and everywhere else, a legacy of the small group of migrants who first stepped out of our native continent and into an uninhabited world. Other great genetic trends, such as those across the New World and the Pacific, track the last migrations into a virgin landscape.

Once, it seemed that modern Europe had a complicated history, with several waves of migration superimposed on each other. The genes of local hunters, who arrived long ago, were—some said—diluted by those of farmers who spread, only a few thousand years ago, from a population explosion in the Middle East. Now, patterns of DNA render it more likely that a great wave of immigrants replaced most of the earlier inhabitants. The archaeology of pots and seeds also suggests that farming was taken up quickly, as soon as people learned about it, with no need for weddings. On the Atlantic coast, the western edge of the new technology, carbon dates taken from charred grains suggest that around 4000 BC farming replaced hunting within only a couple of centuries, too fast for any large-scale mixing of populations. There is no real evidence of a flood of lascivious rustics travelling slowly west across the landscape and mixing their DNA with the continent's existing inhabitants. Instead, the first farmers may have been warriors, for in most places they seem to have replaced their predecessors rather quickly. In Britain, two centres of farming appeared around five thousand years ago, one in southeast England and the other in the south and east of Scotland. Each may have been brought by migrants from what is now France, and the inhabitants of each soon expanded to fill much of the usable land within the British Isles.

As men and women filled the world they killed off many of their kin. The Neanderthals were the first to go. Human habits have not changed since then. Now just a few remnants of our once extensive clan linger on. In a century or so we will be the only large primate (and almost the only large mammal) found outside farms or zoos. Almost all the apes will be extinct in the wild, some before they have properly been studied by science, and much of our biological heritage will be lost forever.

The similarity of some primates to humans had long been noticed by those anxious to judge the evolutionary standing of their fellow men. Charles Kingsley (author of *The Water Babies*) wrote to his wife about a visit to Ireland that "I am haunted by the human chimpanzees I saw . . . to see white chimpanzees is dreadful; if they were black, one would not feel it so much." Chimpanzees are, indeed, our closest living relative. Darwin himself noted that, among their many affinities to humans, they "have a strong taste for tea, coffee, and spirituous liquors: they will also, as I have myself seen, smoke tobacco with pleasure."

Whatever our shared vices, chimps are unlike us in many ways. They are hairy and bad-tempered and do not show the whites of their eyes. They have rather small brains and no ear lobes, and cannot walk upright, float, or cry when upset. They give birth with less pain than we do, and the young mature without any obvious period of adolescence. Our kidneys are better than theirs at keeping salt in the body, and we have more white blood cells. Chimpanzees are in addition safe from the horrors of old age because they tend to die young and, even in zoos, do not get Alzheimer's disease. When faced with illnesses brought on by infection or poor diet, they experience symptoms that may differ from our own (which makes them less useful in medical research than might be hoped).

Chimp sex life has its own flavor. Men lack the penile bone found in male chimpanzees (although when it comes to penis size, man stands alone). Women have outer labia, absent in chimps. Chimpanzee males have larger testes than we do in relation to their body size and, unlike us, seal up their mates with a sticky plug after sex. Promiscuity is the rule. The creatures copulate with enthusiasm (and their close kin the pygmy chimps, or bonobos, are even more energetic). The females show when they are fertile (unlike women, who conceal all signs of that crucial moment), and the males then indulge in a competitive frenzy to mate with them. Sperm from rhesus macaques, a species of monkey known to be very promiscuous, swim faster and lash harder than those of gorillas (in which a single male more or less monopolises the females). Chimpanzee sperm are almost as energetic as those of the macaque, while ours lag well behind both (they do, on the other hand, beat the gorilla's).

The chimpanzee genome was read off in 2005. Not many single letters in the DNA code have changed since the split from our own lineage for, on that simple measure, humans and their closest relative are almost 99 percent the same. At the protein level too, we are close, with about one amino acid in a hundred having altered.

Such figures underestimate the divergence of the two. Changes in the number and position of inserted, repeated, or deleted segments mark both lines. There are three times as many alterations of this kind as single-base changes, which gives an overall difference between men and chimps of around 4 percent. Primates go in for the gain and loss of genes more than do other mammals, and our own lineage is out in front, with a rate of change three times faster than average. As a result, *Homo sapiens* has gained seven hundred gene copies since the split with chimps, and the chimp has lost almost the same number. One chromosome has gone even further. Women have two large X chromosomes, men an X matched with a smaller Y. The human and chimp X have diverged by just half a percent in the DNA alphabet, while the Y has shifted three times more—proof that women, with two Xs, are genetically closer to chimpanzees than are men.

Many of the differences between the two primates have built up because we can modify our environment in a way that they cannot. As a result, we depend less on changes in our DNA than we once did and have lost some genes as a result. Humankind is feebler than it was. We became shaved monkeys with a single mutation (or a few) because a segment of DNA that codes for the hair protein no longer works. It received its fatal blow a quarter of a million years ago. Samson lost his strength with his locks, and so did we, for the DNA behind certain powerful muscles is out of action in ourselves but still active in them (which is why wrestling with a chimpanzee is a mistake). A shared *déjeuner sur l'herbe* is also best avoided, for the animals have enzymes that break down poisons fatal to us. Darwin noted that "savages" ate many foods that were harmful unless cooked (and red kidney beans still fall into that category). Chimpanzees need no kitchens, for they can manage a variety of noxious plants that we cannot. They have also kept many of the talents of taste and smell lost in humans. Many of our own gustatory sensors live on only as battered remnants of once useful structures.

Each species varies to some degree from place to place. The chimpanzee is strongly subdivided. It has three distinct "races," in west, central, and eastern Africa. The central group is about three times as diverse as the western. The extent and pattern of diversity hints that western and central chimpanzees split half a million years ago, while the eastern group separated no more than fifty thousand years before the present day. Humans are tedious by comparison, with far less difference among the peoples of the world than that found among chimp populations.

Sequencing machines are now hard at work on many more of our relatives. Rhesus macaques are small monkeys common in India, Burma, and elsewhere. They are much used in medical research. The animals share rather more than nine-tenths of their DNA with humans. Many of the differences involve—as in the chimpanzee—rearrangements of the order or numbers of copies of particular segments. The animals eat lots of fruit, and genes involved in digesting sugar have multiplied in number when compared with our own. Some of their genes are in a form that in humans leads to disaster. The mutation for the rare inborn disease known as phenylketonuria—a fatal inability to digest certain foods—is the standard version found in macaques. Might some dietary change have rendered lethal an enzyme once useful? A group of genes that predisposes to cancer in humans helps make sperm in macaques. Why does a sexual helper in monkeys cause our cells to run out of control?

Men and chimps, and men and macaques, have changed a lot since their paths parted. Fossils and the genes combine to say when and how their evolutionary divergence, and others from long before, took place. Assuming—and it is a large assumption—that DNA accumulates change at a regular rate, and using well-dated fossils to calibrate the record, the first true mammals evolved around a hundred and twenty-five million years ago. The double helix hints that chimps, orangs, humans, and monkeys cluster together in a class that includes lemurs and rabbits but does not admit horses, dogs, bats, and many other milk-producing animals. The kinship of men, lemurs, rabbits, and the rest is revealed by a certain piece of mobile DNA that hops around the genome. It has been inserted in the same place in all those creatures, proof that they share a common ancestor distinct from that of their furry fellows. Genes, fossils, and geography also combine to suggest that the primates as a group began around eighty-five million years ago. The monkeys and apes split not long after that—which means that their true origin took place as the vast continent of Gondwana was breaking up rather than on the discrete fragments that we now recognize as Africa, Madagascar, and India. The macaque, in turn, set off on its own pathway around twenty-five million years before the present. The split between us and our closer relatives is quite recent. The limited genetic divergence between chimpanzees and humans suggests that they separated five to seven million years ago. Their common ancestor broke away from the gorilla line a million years or more earlier, and that trio split from the orangutan branch about six million years before that.

There is much more to evolution than the random accumulation of error. Darwin's machine may not have a direction, but it can swerve around obstacles as they arise. At the steering wheel is natural selection: inherited differences in the ability to pass on genes. It has led to the striking physical differences between men and chimps, men and macaques, and men and rabbits. Each diverged from a common ancestor, and each has faced its own challenges and found its own set of solutions.

Not all of us will leave descendants, but we all have ancestors. To transmit DNA to the present, every one among them had to survive, to find a mate and to have offspring. Billions more failed at that task. *The Descent of Man* speculates about how selection might have acted upon the human line, although it could offer little direct evidence of its action. Sexual choice was, its author thought, particularly important (and he began, but abandoned, a project to discover whether blondes were less likely to marry than were brunettes); but his case for the role of that process in human evolution was far weaker than the evidence he was able to present for its action in animals and plants.

The evidence for lust as an engine of human evolution is patchy at best, but natural selection of other kinds has without doubt been busy. It leaves its footprints upon the genome in many ways, some obvious and some less so. Often, the evidence of its labours is indirect. Long-term trends such as the increase in human brain size over millions of years show what selection can achieve, given time. The grand patterns of genes across the globe are also proof of its powers.

None is grander than the geographical shifts in man's physical appearance, which are greater than those found in any other mammal. The story of how they evolved has emerged, albeit in several shades of grey, as evidence of how natural selection leads to change and how subtle and unexpected its actions may be. The trends in human skin and hair colour from place to place are due to shifts in the amount of a pigment called melanin. *Homo sapiens* has changed quite recently from white to black, and in some places back to white.

Chimps have rather pale skins (although their faces may become tanned). African skin is, in contrast, dusky, so that darkness is relatively new to the human line. In religious art, Adam and Eve are always shown as light-skinned. Given the look of today's Middle-Easterners, that was unlikely at best. Even in Michelangelo's time people with very fair skin were largely confined to a patch of land within a thousand miles or so of Copenhagen. The first modern humans, a hundred thousand years and more ago, were certainly black.

DNA hints that our swarthy appearance arose well over a million years before the present. At that time, our ancestors had moved from the forests to the sun-baked savannahs. A distinct nose (not found in chimps) also emerged, perhaps to cope with life in the heat of open day. In addition, we lost our hair, probably to cool down. Dark skin was then favoured as it protected against the harmful effects of ultraviolet light.

The first hint of Adam's complexion came not from people but from fish. The zebra fish is much used to study embryonic development. A certain mutant that lacks the dark stripes that give the animal its name is almost transparent. The gene responsible has been tracked down in both its altered and its normal version. A search through our own DNA reveals an almost precise match—so close, indeed, that the human gene reinstates a zebra fish's stripes when injected into a mutant embryo. The enzyme involved shows a large shift in structure across the globe. A certain amino acid is present at one point along the protein chain in 98 percent of Africans, while in 99 percent of Europeans it is replaced by a different version. The form found in Africa makes far more pigment than does its alternative. Much of the difference in appearance between the inhabitants of the two continents hence emerges from a change in a single letter of the genome. The gene varies not at all in its functional section throughout Africa, a hint that dark skin was favoured when it first arose and that later changes have been removed by natural selection. Europeans are far more diverse, with a variety of forms that give rise to black, blonde, or red hair. Other genes further modify skin, hair, and eye colour to give some northern populations skins almost transparent to the rays of the sun. Fossil DNA shows that Neanderthals also had a mutation in the zebra fish gene and that they too were probably white.

The pale skins of China and Japan emerged in a different way. The mutation that lightened the Europeans played no part, for the locals bear the African rather than the European form of that gene. Evolution picked up changes in a separate place in the melanin machine, which, when it goes badly wrong, causes albinism—a complete absence of skin pigment—in Europeans. The loss of melanin from Asian skin comes, in large part, from a less damaging error in that gene. Several other segments of the melanin factory differ in structure between Africa, Asia, and Europe. Most have small but noticeable effects on colour (which is why the children of a marriage between an African and a European vary from dark to light rather than resembling either of their parents exactly).

The earliest modern Europeans and Asians, forty or fifty thousand years ago, were almost certainly black. Even the French cave painters at Lascaux probably had that complexion, for their images of the aurochs, the giant oxen, are reddish, while those of the men who hunt them are much darker. The first Englishmen—those who, thirteen thousand years before today, followed the retreating ice into lands that later became the British Isles—may also have retained some of their African hue when they set foot in their new nation.

Why does it pay to be black in Benin but fair in Folkestone? Almost everything we know about melanin is positive, while fair skin seems at first sight to do more harm than good. Melanin protects against skin cancer. Light skin burns easily—which, although it sounds trivial, makes it hard to sweat and easy to overheat, a situation that brings dangers of its own. In addition, melanin reduces the destruction of vitamins in the blood as they are exposed to the penetrating rays of the sun. Fair-skinned women who sunbathe have reduced levels of folic acid, and their newborn children pay the price, for a shortage of the stuff causes birth defects. Given the disadvantages of pale skin, something powerful must have changed people on their journey from the azure firmament of the tropics to the gloom of northern Europe.

Another vitamin was to blame. Vitamin D helps build bones. It controls the levels of calcium and phosphorus in the blood by helping the gut absorb them and by rescuing quantities of each element that would otherwise be lost in the urine. Oily fish, eggs, and mushrooms are rich in the stuff, and many governments now add it to milk or flour to promote their citizens' health. Vitamin D can, in addition, be made in the skin through the action of ultraviolet light on a form of cholesterol.

To do the job, the light must get in; and melanin keeps it out. Africans have to spend several hours a day in bright sunlight to make enough vitamin D to stay healthy, but northern Europeans who expose their arms, head, and shoulders for fifteen minutes in a midsummer noon can make enough to meet their needs. Without vitamin D, children are in danger of contracting rickets, which leads to a curved spine or legs and can cause severe disability. Rickets may also cause seizures and spasms—a side effect of calcium shortage—which can end in heart failure. Nine out of ten infants in Victoria's smoky and starved cities showed signs of the illness, and rickets is still the commonest noninfectious childhood disease in the world.

Most young black people in the United States have low levels of vitamin D, and rickets is, as a result, three times more common among black Americans than in their white fellow-citizens. The athlete O. J. Simpson suffered from it as a boy and wore homemade leg braces. In England, my own generation was saved by cod-liver oil distributed to citizens free of charge, but the modern world is not so lucky. Rickets is back in Great Britain, and a third of Asian and Afro-Caribbean children are short of the crucial vitamin (for the former the ban on uncovering themselves is much to blame). Severe deficiency is nine times more frequent in that group than in Europeans, and one in a hundred of their children show signs of illness. Girls do worse, which is bad news later on, for the condition causes the pelvis to narrow, resulting in difficulties during childbirth. Vitamin D deficiency has even been seen in affluent white children allowed to play in the sunshine—but protected from ultraviolet rays with sunscreen.

The magic substance also helps to hold diabetes, arthritis, muscular dystrophy, and heart disease at bay and protects against certain cancers (which may be why cloudy places show higher rates of lung and bowel cancer). Any change in skin colour that helped to generate more vitamin D must have been most helpful on mankind's journey into the gloom. Natural selection favoured the new mutations at once, and under grey skies fair skin took over.

Selection has led to other upheavals in human DNA. Many emerge from changes in our habits—migration to a place with a different climate, shifts in diet, and the development of towns and cities which led to outbreaks of new infectious diseases. For nine-tenths of our history as a species, most of mankind saw fewer people in a lifetime than an average Westerner now encounters on his or her way to work. Farming led to a population explosion, and *Homo sapiens* is now ten thousand times more abundant than any other mammal of a similar size. In a world of pathogens and parasites, abundance is an expensive luxury.

Epidemics have often cut our species down to size. They need large populations to sustain themselves, and migrants to spread the infection. The Plague of Justinian, which began in Constantinople in AD 541, killed off a quarter of the people of the eastern Mediterranean. The Black Death spread along the Silk Road from China in the fourteenth century and returned again and again to the teeming and filthy cities of the west. Two out of three Europeans died. Sickness is potent fuel for selection.

One illness shows its power better than any other. A third of the world's population is exposed to malaria, half a billion are infected, and the disease kills five people a minute. The real attack began only ten thousand years ago, when men opened clearings in tropical forests at a time of warm, wet weather. That helped mosquitoes to breed and the parasite to spread.

In Kenyan families, poor conditions—a marshy spot, too much rain, too many children—explain some of the variation in individual risk of infection, but genetic differences are behind a third of the overall chance of ending up in the hospital. Some variants have a large influence and are soon picked up by evolution, while others are more subtle. The most important involve changes in the red blood pigment, haemoglobin. A quarter of a billion people bear at least one copy of one or another such genes. The best known is sickle cell, a single-letter change in the DNA alphabet. The haemoglobin of those with two copies of that gene forms long crystals in parts of the body low in oxygen. The red cells take up a crescent shape that restricts circulation and causes pain, heart disease, and worse. Those with a single dose of the altered message—a third of all Africans—are healthy. They have half the risk of developing a fever if infected with malaria and a tenth the risk of becoming seriously ill. The altered gene is also common in southern Europe, the Middle East, and India. It has arisen on at least four different occasions. Other single-letter changes, giving lesser protection, are found in countries such as Bangladesh, while deletions of long or short sections of DNA do the same in the Middle East and Oceania. Once again, those who bear a double dose of a damaged gene pay a severe price, while people with just one are safe from infection.

Many other mutations have been pressed into service against malaria. The parasite hijacks the metabolic enzymes within our own red cells to fuel its own machinery. Hundreds of millions of people bear a defective version of one of those enzymes and in return gain protection. One form of the parasite cannot get into cells that lack a certain attachment site. Almost all West Africans have such cells. Elsewhere, a change in the shape of the red cell baffles the infection. The high salt and iron levels in African blood also fend it off. Dozens of sections of DNA are implicated in the fight against the agent of malaria and many no doubt remain to be discovered. Large or small, each of these variants has been propagated by natural selection, which, as always, cobbles together a response with whatever mutations turn up.

Natural selection is poised to deal with enemies as they arise. Wherever it works, it leaves evidence, often indirect, that it has passed by. Some changes in DNA alter the structure of proteins while others do not. The ratio between the two kinds of mutation is a crude gauge of its actions, for a shift in protein structure is more likely to be visible to selection than is an alteration of the

DNA code that causes no such change. On that criterion, our lineage has experienced considerably less selection than has that of the chimpanzee since our families parted.

Another clue to the action of Darwin's agent comes from the blocks of genetic variants packed close to each other along each chromosome. As a favoured gene—perhaps a new antimalaria mutation or one for a change in skin colour—is picked up and becomes more common, it will drag along sections of DNA on either side. The stronger and more recent the selection, the longer the accompanying segment. In Africa the sickle cell variant sits in the middle of great sections of the double helix that scarcely vary from person to person, while the same is true for the pale skin gene in Europe. That pattern hints that the new mutations were immediately seized upon and spread—with their neighbours—fast through the population.

In time such uniform blocks of DNA are broken up by the random shuffling of genes that takes place when sperm and egg are formed, but that process may take a long time. A length of DNA that varies scarcely at all from person to person, within the generally diverse genome, is hence evidence that selection has been at work not long ago. The human and chimp genomes each have thousands of such segments. One gene in sixty among the chimps bears that Darwinian mark but only half as many in humans—further proof that we have been able to cope with new challenges in a manner they cannot. Man's ability to modify the environment has, it seems, weakened the hammer blows of nature. Antimalaria drugs now do what once could be achieved only by expensive mutations, and while, thousands of years ago, our skin responded to a shift in climate with a genetic change, today most people, black or white, protect themselves against the sun with clothes.

The loss of our native nudity was an early hint of that unique evolutionary talent—to respond to a challenge not with bodies but with brains. This habit allowed us to spread across the world, for with clothes we take the tropics with us wherever we go.

Adam and Eve, in their sultry paradise, were unashamed, but after the first (and least original) of all sins they made aprons to hide their nether parts. When did they put them on? Lice hint at when clothing was invented. Chimps and gorillas have lots of those fellow travellers and spend much time in grooming to get rid of them. When humans emerged onto the sunny savannahs they lost their hair. The lice had a hard time and evolved to live in the only patches of habitat left. We are now hosts to three distinct kinds, head, body, and pubic lice, with the body louse the only one able to hang on to clothing. The pubic version is closest to the lice found on gorillas and may have joined us from there. DNA shows that the other two evolved from a chimpanzee parasite that began its intimate acquaintance with ourselves six million years ago and that the body and head forms separated no more than fifty thousand years ago. Their split may mark the moment when we first donned garments and gave an enterprising insect a new place to live. Our parasites hint that men dressed up as they took their first steps towards the icy north.

Since then we have learned to cope with such visitors through insecticides, to deal with cold with central heating and with noxious foods with kitchens. Our talents emerge from the contents of the skull. Much of what the brain can do is—like Adam's underpants—unique. Darwin noted that "there can be no doubt that the difference between the mind of the lowest man and that of the highest animal is immense." To understand human evolution we need to know how and why that most human of organs is so different from that of any other primate, and human behaviour even more so.

The brain is three times as big, and the cortex, the thoughtful bit, five times larger than that of the chimpanzee. To accommodate it, the modern skull has become several times roomier than that of three million years ago. Chimps are born with a brain almost as big as that of an adult. Human babies, in contrast (whose brains are already larger than that of a chimpanzee), continue to invest in grey matter until they are two. Genes active within the cranium have multiplied themselves in comparison to those in other primates. One such gene, which, when it goes wrong, leads to the birth of infants with tiny heads, has evolved particularly fast. The nerves within the human skull are more interconnected, and the junctions more sophisticated, than those of the chimp, and the structure is also busier at the molecular level. Even so, much of the DNA most active in that part of the body has changed no faster than that at work in liver, muscle, or scrotum.

However it works, the brain is expensive. Pound for pound, it burns up sixteen times more energy than our muscles use, a quarter of the resting body's budget. The chimpanzee brain uses only half as much. How can we afford such a luxurious appendage? We consume no more than other primates of our size, but humans have a richer diet, with more meat and fewer roots and leaves, than that of our relatives. Cooking, too, does a lot to extract the goodness from food before we eat it. As a result we need smaller intestines to soak up nutriments. Humans also invest less in muscle than do other apes. All this began, like black skin, a million and more years ago, when we moved from forests to savannahs, travelled in larger groups, and became better hunters with a meatier diet and, in time, learned to become primitive chefs. The way to man's brain was through his guts.

Be that as it may, today's organ of thought is no bigger than a Neanderthal's. Fossils of their newborns show that they were born with a brain as large as our own, which grew even faster during infancy. Even so, those creatures acted far more like apes than we do. Something more than an extra dose of grey matter has made us what we are. To quote Darwin: "Of all the differences between man and the lower animals, the moral sense or conscience is by far the most important." A glance at our relatives shows how right he was.

Chimpanzees are nastier than many people like to think. They kill monkeys and are pretty unpleasant to each other, too. Their sex lives would shock Queen Victoria, and their ethical universe, if they have such a thing, is much darker than our own. They live in groups, but the groups break and reform as their members quarrel. Terror makes their world go round. If two chimps need to pull a rope to get a tray of food, they will, but only if they are out of reach of one another. Otherwise, the dominant animal attacks its subordinate and neither then gets anything. Anger and greed destroy the hope of reward. What makes humans different is a loss of fear, odd as that sounds in a world where that emotion seems to be everywhere. When anxiety goes, society can emerge.

Our social skills begin early. A group of two-year-olds asked to find a piece of food after watching it being moved or turned round, or put in a box that makes a bleeping noise were pretty good at each job—but no better than adult chimpanzees, for both babies and chimps succeeded at about two trials out of three. When they were able to learn from others the babies won hands down, for they became far better at each problem when they saw someone else solve it, or when an adult pointed to, or looked towards, the hiding place or made noises that told the infants that they were getting close to the hidden food. To use such clues calls for insight into another's feelings. We have much more of that talent than do our relatives. The chimps took no notice of those trying to help them and stayed hungry as a result.

Chimpanzees learn, but they do not teach: they ape but do not educate. In some places, adults fish for insects with a stick or bash nuts with a stone, and the young emulate them; but the grown-ups make no effort to show the infants how to do the job, do not change their ways when youngsters are around, and never check to see how well they are doing. Even birds, with their bird brains, can do what a chimp does. A budgie will pull out the stopper of a bottle of food if it sees another do it. Crows, in turn, are even smarter.

A chimpanzee parent's negligence about the next generation is a reminder that the minds of our hairy relatives are less like our own than we might imagine. Real education asks for more than mere copying. A good pedagogue can instruct his charges about anything as long as he keeps a few pages ahead and they respond to his efforts. Teachers have insight into the mental lives of their pupils, into who is doing well and who is struggling, and know how to encourage them and keep them from growing bored. A competent teacher needs to understand what his students are thinking—and chimps do not: they have no "theory of mind," as psychologists put it. We have lots, and it helps those on both sides of the lectern. Much as most teenagers might doubt it, no ape could ever become a schoolmaster.

The best way of reading a mind is to chat to it. Thomas Love Peacock invented a character called Sir Oran Haut-Ton, who learns to play the French horn but not to speak (he is elected to Parliament, where his silence gives him an air of wisdom). *Homo sapiens* is the eloquent ape. Speech is the scaffold upon which society is built. Even a group of deaf children will babble with their hands. No other primate can do anything like that, and all attempts to persuade them to do so have failed (Noam Chomsky noted that it was "about as likely that an ape will prove to have a language ability as [that] there is an island somewhere with flightless birds waiting for humans to teach them to fly.")

The origin of language is a cause of endless dispute that, given that just one creature can speak, may never be resolved. *The Descent of Man* suggests that language started with love songs and speech was a side effect of sexual selection or, less romantically, "Some unusually wise ape-like animal should have thought of imitating the growl of a beast of prey, so as to indicate to his fellow monkeys the nature of the expected danger." Perhaps it did; or perhaps it grew instead from the simple fact that we are social creatures. Apes groom each other because the constant pacification calms them down and reduces the conflict that is never far from the surface. Big groups demand too much scratching time, but reassuring sounds can placate lots of individuals at once. The savage breast might first have been charmed in that way, perhaps, indeed with song (which might be why some stutterers can sing a sentence that they otherwise could not speak).

However it began, language makes us what we are. The ability to speak resides in the left side of the brain. Plenty of primates have brains almost as lopsided as ours. Even so, chimp tongues fill their mouths while ours are comparatively dainty. The human tongue has retreated down the throat. The language of Shakespeare is a complex set of sounds made as the space above the larynx flexes and bends. The anatomical changes leave evidence in the shape of the skull. Neanderthals had chimplike mouths and could probably do little more than grunt. The first throat and mouth capable of speech may have emerged no earlier than fifty thousand years ago—not long before the explosion of technology that led to the modern world.

One child in twenty has some form of speech disorder. One rare genetic abnormality makes it impossible for those who inherit it to cope with grammar. Baby mice with the same damaged gene make fewer squeaks than usual when removed from their mothers, and people with a version impaired in a different way are at risk of schizophrenia; of, like Saint Joan, hearing voices that are not there. The normal version found in humans differs in two of its amino acid building blocks from that in all other primates. The gene involved helps in muscular coordination and does many other things. Its role in the ability to speak may be indirect, for a variety of other parts of the double helix—a rather different set in humans and chimps—are under its control, and they may be the main players in the emergence of our unique talent. It is foolish to speak of a gene for language, but if the transition from animal to human turned on speech it might have involved rather few molecular changes. The situation is confused by the discovery that Neanderthals had the human version of the gene, which must hence date back to our inarticulate joint ancestor.

Wherever they came from, words are the raw material of a new kind of inheritance in which information passes through mouths and ears as much as through eggs and sperm. It moved us on from being a rare East African ape to the most abundant of all mammals. Ideas, not genes, made us human. Our DNA is not much different from those of our kin, but what we do—or say—with it has formed our fate.

Charles Darwin's famous "indelible stamp" is, without doubt, firmly imprinted into the human frame. Modern biology shows that chimpanzees are even more like us than he imagined—but only in the most literal way. The strengths (and the limitations) of evolutionary ideas in understanding what makes us human have grown ever more clear. His theory is powerful indeed but we need to be reminded where its competence ends.

In 1926, the Soviet government sent an expedition to Africa. It was directed by Il'ya Ivanovich Ivanov, famous for hybridising horses and zebras by artificial insemination. The Politburo hoped to do the same with men and apes, in an experiment that would be "a decisive blow to religious teachings, and may be aptly used in our propaganda and in our struggle for the liberation of working people from the power of the Church." In Guinea, Ivanov obtained sperm from an anonymous African and inseminated three chimps—but none became pregnant. He then planned to fertilise women with chimpanzee sperm, but was prevented from so doing. Back in Russia, he set out to do the same with a male orang-utan and a woman who had written that "with my private life in ruins, I don't see any sense in my further existence. . . . But when I think that I could do a service for science, I feel enough courage to contact you. I beg you, don't refuse me . . . I ask you to accept me for the experiment." The orang, alas, died before its moment of glory and Ivanov was arrested and exiled to Kazakhstan, where he, too, met a childless end.

Americans anxious to stop research in human genetics once attempted to patent the idea of a human-chimp hybrid in the hope of whipping up protest. The application was denied on the ground that the U.S. Constitution does not allow the ownership of human beings (whether the cross-breed would have that status was not discussed). Artificial fertilisation of a chimpanzee egg with a man's sperm may now be feasible (although claims to have produced a "humanzee" are fraudulent), but is universally seen as beyond the pale. The problem is not one of biology but of what it means to be human. A hybrid between a chimp and *Homo sapiens* makes too ready an equation between our apish bodies and our immortal minds.

Charles Darwin was well aware of the limits of his theory. As he points out in the famous last sentence of *The Descent of Man,* men and women possess noble qualities, sympathy for the debased, benevolence to the humblest, and an intellect which penetrates the solar system. All that does not change the fact than in our bodily frames, most of all when reduced to chemical fragments, we bear the indestructible mark of our humble ancestry.

Some people despise his science because it appears to destroy man's special place in nature; but they misunderstand what evolution is all about. The double helix does not diminish *Homo sapiens* but sets him apart, on a mental and moral peak of his own. The theory of evolution does not render us less human than we were. Instead the insight it provides into man's place in nature shows us to be far more human than we ever realised. A century and a half after Queen Victoria's disagreeable visit to Jenny the Orang-Utan, I gave a talk at London Zoo which pointed this out—and most of the apes agreed.

CHAPTER 2

The Green Tyrannosaurs

SOARING ABOVE SOUTHERN VENEZUELA IS A HIDDEN LANDSCAPE, the sandstone plateau of Mount Roraima, an inaccessible peak often shrouded in mist. Arthur Conan Doyle used the place, or one much like it, as the location for his 1912 book *The Lost World*, a tale set in a land of evolutionary imagination—a place of dinosaurs, ape-men, and primitive humans, ready to be explored by the irascible Professor Challenger. It was a fearsome spot, but the bearded Englishman lambasted the lizards and saved the savages, just as any Edwardian reader would expect.

Conan Doyle was born in the year of *The Origin*. By his fifty-third birthday, the theory of evolution had become so widely accepted that a literary hack could use it as the centrepiece of a work of fiction. Doyle (who had read the reports of the British explorer who discovered the unique island in the sky) seized the chance, and his book sold hundreds of thousands of copies to a well-primed public.

In reality the dinosaurs had disappeared from Roraima millions of years earlier, and the local "savages" never made it to the top. Even so, its remote summit is a genuine lost world, not of giant anthropophagous lizards but of unassuming plants with similar dietary habits. Those green carnivores eat not human flesh but insects. They must do so or starve.

Their practice is widespread. Almost six hundred insect-eating plants, from all over the world and from a wide variety of groups, have been discovered. Such a way of life has evolved independently many times, and the tactics used to trap and digest prey are varied indeed. Separate lineages, starting from different places, have taken up an identical diet. They have come to solutions that at first sight seem remarkably similar when it comes to the need to find, digest, and absorb food. Charles Darwin had used such convergent evolution, as the process is known, as evidence for natural selection in *The Origin of Species*. The physical resemblance of Australian marsupials such as the Tasmanian wolf to the mammalian wolves elsewhere in the world, or of wings in birds and bats, was, he saw, powerful proof of its action. Different creatures faced with the same challenges adopt structures and habits that look much the same but have very different roots. Life can, as a result, reach what seems an identical end by quite distinct pathways: as he put it in *The Origin*, "In nearly the same way as two men have sometimes independently hit on the very same invention, so natural selection, working for the good of each being and taking advantage of analogous variations, has sometimes modified in very nearly the same manner two parts in two organic beings, which owe but little of their structure in common to inheritance from the same ancestor."

We know many such examples—flight, not just in birds and bats but in squids, fish, dinosaurs, flying squirrels, and the marsupial sugar-glider of Australia (not to speak of the flying snake whose flattened body allows it to glide from tall trees). We ourselves are not immune to convergence, for plenty of creatures have lost their hair, grown their brains, or even (like the meerkats of Africa, who instruct their infants how to eat poisonous insects) stood upright and gained some simulacrum of culture.

Convergent evolution in response to a common challenge is almost universal. It has been so effective that it can disguise the real shape of family trees. Many natural pedigrees have now been revealed with DNA. Sometimes it shows that creatures thought to be close relatives are in fact not kin: thus, the vultures of the Old and New World, similar as they appear, do not have a recent common ancestor. The former are eagles and the latter storks. Anteaters and aardvarks, lions and tigers, moles and mole-rats—all hide a bastard ancestry beneath their shared appearance. The process goes further: on Roraima itself, for unknown reasons, melanism is rife and the rocks harbour black lizards, black frogs, and black butterflies. The mutation responsible is the same, or almost so, in zebra-fish, people, mice, bears, geese, and Arctic skuas (and perhaps in lizards and frogs) and has been picked up by natural selection in each. On a more intimate scale, the complicated chemical used as a sexual scent by certain species of butterfly does the same job for elephants (which is riskier for one partner in the relationship than for the other). Within the cell, too, shared evolutionary pressures have produced enzymes with distinct histories that have settled on an almost identical DNA sequence in their active parts.

Evolution towards a common plan is also rife among plants. The cactuses of the Americas—spiny, thick-skinned, and globular—resemble the euphorbias of South Africa, with which they have no more than a distant affinity.

Just after the publication of *The Origin*, Darwin began to study another botanical lifestyle that drags a great diversity of unrelated species into a shared set of habits. His interest began in 1860 when he visited Hartfield in Sussex, the home of his sister-in-law Sarah Elizabeth Wedgwood (and later the birthplace of Winnie the Pooh). There he saw thousands of sundews—small clumped plants with a sticky surface that traps insects. Some had as many as thirteen victims on a single leaf. Most were small flies, but the prey included large butterflies, and he was told that the traps could even catch dragonflies. As so many sundews were present, the numbers of insects slaughtered must, he thought, be prodigious. Each leaf had scores of glands held upright on fine hairs. They exuded glistening globules of liquid even on dry days and entangled any small creature foolish enough to land upon them. The sundew, he found, had feeble roots—evidence that most of its nutrition came from its grisly way of life. Darwin brought some specimens back to his greenhouse and began to explore how they did their job. It was the first step in a decade of work that produced a powerful justification of his claim in *The Origin* that natural selection could, starting from separate places, end up with much the same result.

In 1875 he published *Insectivorous Plants*, a book of almost five hundred pages. It deals not just with sundews but with a variety of such creatures from around the world (some from the area of Roraima itself). Darwin found many independently evolved similarities among the very different plants that had taken up the habit. A closer look showed that many of their adaptations are shared with animals. In reaction to her husband's work on one particular insectivore, Emma Darwin noted in her diary, "I suppose he hopes to end in proving it to be an animal." Her marital partner was so astonished by the parallels he found between the two kingdoms that he wrote to a friend, "I am frightened and astounded at my results." The aesthete John Ruskin was less impressed, as he wrote: "With these obscene processes and prurient apparitions the gentle and happy scholar of flowers has nothing whatever to do. I am amazed and saddened, more than I can care to say, by finding how much that is abominable may be discovered by an ill-taught curiosity." Darwin's curiosity, ill-taught or not, added another plank to his evolutionary edifice.

Roraima and the flat islands of rock around it are ancient indeed. Its sandstone peak—once part of a wide and barren plain, most of it eroded away—is almost two billion years old. In the context of its immense history, the terrible lizards left not long ago and the humans clustered around its base arrived in an evolutionary yesterday. Its unique vegetation has been on its rain-soaked flanks for longer than either. The plants have seen the slow passage of time and have themselves changed to match. One in three evolved upon the mountain's lonely rocks and is found only there. Their native land is a hungry place. Constant downpours eat at the soil and strip off its nutrients, tipping them down some of the highest waterfalls in the world. The rain washes away the nitrogen that every tree, shrub, and flower needs to grow. Sandstone peaks, deserts, dunes, bogs, pine forests, Mediterranean scrublands, and more—all are short of that element, and each has evolved a distinct set of inhabitants whose battle for existence is focussed on the need to find it. The struggle for nitrogen shows—better even than the wing or the eye—how natural selection can reach the same end with different means in creatures from entirely unrelated parts of the biological universe. Plants, animals, bacteria, and fungi are drawn together in their shared craving for a simple nutriment, and all have become entangled with each other in their struggle to find it.

The gas makes up four-fifths of the air but plants cannot extract it directly. Their growth is often limited by a shortage of the stuff. Many manage by soaking it up from the soil. They forage like hungry animals with their roots, which stretch further and further as the essential item runs short.

Places such as Roraima do not have enough nitrogen to allow plants to survive even with such help. They are forced into contracts with other creatures that donate the vital element. Sometimes the treaty seems positive, although in truth it is based on greed and expedience. In other cases the divergence of interests between the parties is clear. Plants starved of nitrogen eat insects and soak up the element held in their flesh. The trade is one-way: the plants kill the animals and the latter get nothing in return.

All animals fall prey to the vegetable world in the end as their dust returns to dust, but in the starved landscape of the Lost World the vegetation cuts out the middleman and devours the local wildlife directly. The habit has evolved all over the world. Natural selection has tinkered with leaves, roots, and other parts to generate the equivalents of teeth, gullets, stomachs, and intestines. As it does, it has drawn some of the machinery of the botanical world close to that of animals.

Charles Darwin's work on the insect-eaters sold less well than had Conan Doyle's book on imaginary lizards, even though his subjects behave in a way beyond the imagination of the most fanciful novelist. *Insectivorous Plants* raised biological questions that resonate far beyond the universe of carnivorous plants. At first its author doubted the value of his own work and wrote to a colleague: "I must consult you some time whether my 'twaddle' is worth communicating," but he soon became an enthusiast. His volume on those creatures is a masterful narrative that spells out the ingenuity of existence.

Plants that eat flesh had attracted interest long before Darwin's day (they still do, although the Australian carnivore that feeds on rabbits has yet to be confirmed by science). The sundews he saw in Sussex have a natural flypaper that holds its victims with a syrupy glue. The leaves curl round to entangle them before they meet a sticky end in its sinister digestive globules. They give it a name; as Henry Lyte wrote in 1578 in his *Nievve Herbal*—"This herbe is of a very strange nature and marvelous: for although that the Sonne do shine hoate, and a long time thereon, yet you shall finde it always moyst and bedewed." *Insectivorous Plants* also describes experiments on the Venus flytrap, which modified some of its leaves into a "horrid prison," and on many other specimens forwarded to Down House from afar.

The insectivore habit has evolved in around a dozen lineages, and the sage of Downe saw most of them. They represent but a small fraction of the quarter of a million kinds of flowering plants, but their habits, and their origins, are varied indeed. Some of the flesh-eaters are as closely related as are, say, the anteater and the pangolin (long-nosed insectivorous mammals from the Americas and the Old World, the former close to armadillos and the latter to dogs and cats), but others are as different as an anteater is from an insect-eating lizard or a swallow. Natural selection has caused widely separated branches of the tree of life to turn to the same expedient.

The sundew belongs to a group of a hundred or so species whose centre of diversity is in Australia. It has a relative called the butterwort that hunts in the same way. The flypaper habit has evolved on at least five independent occasions, to produce the Australian "rainbow plants" (named for the sinister sheen of their leaves) and more. Some are ten feet high while others are tiny, and they are found around the globe, from Alaska to New Zealand (although Europe has only three of the hundreds of species known). Even within each group there is diversity, for DNA shows that the sundew and a similar species from Portugal evolved insectivory of their own accord, as each can trace affinity to closer relatives that do not indulge in the pastime.

A second trick is to swing a door shut upon the prey. The most familiar jailer is the Venus flytrap. In 1768 Arthur Dobbs, governor of North Carolina, sent the first specimen of the "fly trap sensitive" back to Britain. The botanist William Bartram saw the "ludicrous" plant on his travels through the Carolinas: "This wonderful plant seems to be distinguished in the creation, by the Author of nature, with faculties eminently superior to every other vegetable production. . . . We see here, in this plant, motion and volition." The "irritable principle in vegetables" was much commented on, even if some denied that any plant would lower itself to prey on an animal. Linnaeus, the great classifier, insisted that the flytrap always let its prisoners go, and others thought the insects trapped within were sheltering from frogs. This idea became part of social theory. The anthropologist César Lombroso (who believed that crime was a biological throwback beyond the control of those responsible) saw flytraps as marking "the dawn of criminality." They "establish that premeditation, ambush, killing for greed, and, to a certain extent, decision-making (refusing to kill insects that are too small) are derived completely from histology or the microstructure of organic tissue—and not from an alleged will."

The flytrap (only one species is known) lives in nitrogen-poor swamps in the Carolinas. It was given the erroneous Latin tag *Dionaea muscipula* (which means "mouse-eater" rather than "fly-eater" as intended), but the colonists gave it the name "tipitiwitchet," then a slang term for female genitalia, because of its supposed resemblance to that organ. (To avoid vulgarity, Thomas Jefferson used the label "Aphrodite's mousetrap" when he added a specimen to his collection.) The traps are modified leaves. Each plant bears up to a dozen. Darwin himself considered the flytrap to be "one of the most wonderful in the world." Although it did not glue insects to its leaves but enclosed its prey with spines that interlocked like the teeth of a rat-trap, its sensitivity reminded him of the sundew's entirely distinctive hunting strategy.

Just two snap-traps are known; the Venus flytrap and the unrelated waterwheel plant (again just a single species, but found scattered across the world), which does the same under water, on a smaller scale. Other freshwater flesh-eaters use a different method involving a lobster pot, a snare with a one-way entrance valve. A separate group, found both on land and in fresh water, has tiny capsules or bladders which, when touched, suck in prey with irresistible force. The bladderworts, as they are called, have hundreds of species, found everywhere except in Antarctica. They prime each trap by pumping water out across its wall. Another kind, found on moist rocks in South America and parts of Africa, shows a microscopic kind of carnivory, ingesting protozoa that swim into tiny slits in its subterranean leaves. Darwin had speculated that it was indeed a meat-eater, although he had no idea of its food.

The insect-killers of Roraima use a different trick, a pitfall based on rolled leaves with margins sealed together. The fatal ambush is covered with a slippery glaze and decorated with a nectar bait. The sixteen species known from that peak are relatives of heathers and their mountain allies. Some are a metre tall while some are tiny. Other pitcher plants of the New and Old Worlds and of Australia also use rolled-up leaves to snare their prey (and some are big enough to take mice), although they come from another section of the botanical kingdom. Their capital is the New World, which has more of those baleful creatures than anywhere else.

The cobra lily of the western United States has a mouth said to resemble that of a snake. It feasts on ants. Others among its kin have a flared cover that shelters the opening of the trap and stops it from filling with water. A second group with the same general appearance, the monkey cups from around the Indian Ocean, make their pitfalls as structures that spring from a leaf's mid-rib and are held at the end of a long tendril. They get their Latin name, *Nepenthes*, from the grief-dispelling drug given to Helen of Troy. Linnaeus was impressed by them: "What botanist would not be filled with admiration if, after a long journey, he should find this wonderful plant. In his astonishment, past ills would be forgotten when beholding this admirable work of the Creator!" Yet another pitcher is found in western Australia. This plant, whose traps look rather like old shoes, is taxonomically a long way away from any of the other pitfall-makers.

DNA reveals some unexpected affinities among the pitcher plants. The Old World kinds are more closely related to sundews and Venus flytraps than to their New World equivalents. In addition they are quite close to non-carnivorous tropical lianas, and find more relatives in a larger class that includes rhubarbs, spinach, and beets.

An even more distinct group, the bromeliads—relatives of the pineapple and in quite a different subdivision of the plant kingdom from the other green carnivores—also make leafy containers that fill with water. They live in the tropical forests of the Americas. There may be more than a hundred thousand in every hectare, most of them growing upon trees. The vessels made by their fused leaves generate a huge series of tiny lakes, in which a variety of different creatures find a home. At first the bromeliads seem benign, for they lack the digestive enzymes found in the other pitfalls. Tadpoles, insect larvae, inch-long salamanders, and tiny crabs all live in the liquid within. In truth they have a darker side. Each watery island is full of conflict, and their proprietors gain nitrogen from the corpses of the creatures that are killed there and broken down by bacteria. Only one kind has taken a step towards true carnivory by making digestive enzymes of its own.

Still other meat-eaters, unknown in the nineteenth century, are bizarre indeed. Certain soil fungi devour nematodes, wormlike creatures far larger than themselves, with a lasso that snaps shut within a tenth of a second, strangles the animal, and gives the predator a rich source of food. Others do the same job with a sticky pad, while the familiar (and tasty) ink cap mushroom puts out spiked and lethal balls that puncture its prey and allow the fungal spores to grow within it. A hundred and fifty fungi that snack on living flesh are known, and there is a whole world of animal-hunting mushrooms waiting to be discovered.

Wherever they sit in the botanical world, life in a hungry place has pushed every flesh-eater towards a similar set of expedients. Like cactuses they succeed where others fail, in their case because of a shortage of food rather than of water. The cost of success is specialisation. Their habit can be expensive and their way of life fragile, for traps are hard to make and often impose a reduced investment in roots or leaves. Insectivores hence find it difficult to cohabit with other species (which means that they carpet vast and exclusive areas), are forced to seek sunlight because their leaves are too feeble to cope with shade, and do best in places where fires occasionally sweep through and wipe out the competition.

The insect-eating package may be lost when conditions change. Some species are committed to it. The Venus flytrap is trapped in its narrow way of life, for it gains three quarters of its nitrogen from insects, while the cobra lily is not far behind. Others are less so. Most such plants have some chlorophyll, the stuff that makes leaves green, but often no more than half that found in normal species. They gain some energy from the sun, though at reduced efficiency. The sundew and many of its fellows have reduced roots, as not much food is available in their native swamps, but they can soak up a little. Radioactive labelling shows that up to half of the nitrogen taken up by a typical individual comes from soil rather than from flesh.

For all of them, prey are more important during the summer when they are abundant—and a large part of what they provide goes to make flowers, expensive as they are. The bladderwort makes traps only at the height of the season, when swimming food is common and the time has come for sex. The pitchers of New England make more traps in the bogs most lacking in nitrogen, but put effort into ordinary leaves when the water contains more of that element. Pitchers make two kinds of leaves, either modified to make a trap or large and flat to soak up sunshine. When nitrogen is present, the plants make more of the latter. Others, too, play the commodity market for the sundew produces less slime than normal when given a decent dose of fertiliser. All this suggests that carnivory is a last resort, abandoned whenever a cheaper source of the essential ingredient becomes available. A return to vegetarianism has happened many times, for DNA shows that several plants that evolved from meat-eating ancestors now have a more conventional diet. Thus, certain pitchers in Borneo soak up nutrition from dead leaves or from bird droppings that fall into their flasks instead of from insects.

The first botanical carnivores evolved long ago. A famous fossil bed at Yixian, in northeast China, from around a hundred and twenty-five million years ago, yields dinosaurs with feathers—together with a small pitcher quite similar to those of today, with lures to attract its prey and glands ready to soak up their remains. The bed dates from around the origin of flowers themselves. A family tree of the flowering plants based on DNA suggests that perhaps sticky traps evolved first, while pitfalls and snap-traps came later—which, if true, pushes the habit even further into the past. Even the nematode-eating fungi have left a hundred-million-year-old fossil, trapped in amber in a French quarry.

Darwin transformed our insight into that motley set of unrelated creatures, for he raised—and answered—biological questions that resonate far beyond their own narrow universe. He began a systematic survey of how the group's various members caught their prey, digested it, and absorbed its goodness. The results, he thought, were "highly remarkable."

First, he found that sundews grew far better when fed insects than when starved (although they could survive for a time on a normal diet). Their abilities were impressive. Even a tiny gnat with its "excessively delicate feet" set off a reaction. A single gland moved within ten seconds of being presented with a meal and the impulse to do so soon spread through a whole leaf. Within an hour of the prey's arrival, a mass of tentacles began to bend towards it. Light and dark made no difference, although the trap moved faster on warm days. The tentacles had no sense of smell—an object had to touch the surface to induce the effect—but they could taste. They held on to pieces of meat longer than to bits of glass, cork, or hair. Water did nothing to excite them, nor did a firm prod with a twig, but even the most minute particle of living material led to a response.

What woke up the sundew was nitrogen. Urine (presumably the great naturalist's own) did the job, while tea did not. Ammonium carbonate—*sal volatile*, used to stimulate fainting Victorian ladies—contains the essential nutriment and elicited a response almost at once, even when diluted. Just twenty-millionths of a grain of nitrogen salts in solution—less than a thousandth of a milligram—was enough to stir the sundew's interest.

The Venus flytrap solves the problem of living off insects with a different set of tricks. Its prison doors slam shut within a tenth of a second of being touched—among the fastest of all plant movements. The flytrap's trigger—which evolved quite independently from that of the sundew—gives further hints about how information is translated into action. Darwin found the flytrap as alert to a sudden touch as was the gland of its sticky fellow-carnivore, but it responded less to prolonged pressure. Two or three quick taps within a span of thirty seconds, rather than a single one, were needed to spring it, perhaps to avoid accidental disturbance by wind-blown grains of dust. Rain had no effect.

As soon as the prey arrives, it disturbs the sensitive hairs that cover the trap. A wave of electrical activity passes across the surface at a rate of around four inches a second. The wall of each modified leaf, with its two hinged parts, has cells kept full of liquid at high pressure, and the flytrap uses its feeble powers to pump them up, with many hours needed to reset the snare after each use. The trap wall is curved outward when open and inward when shut, with a fine balance between the two stable states. A squeeze with the fingers causes it to snap shut in much the same way as a pea pod pops open. The elastic energy stored in the curved shell of an open trap is released in a sudden rush and it slams closed.

Pitcher plants use a different trick to catch their prey. Many befuddle insects with signals that promise food or sex but that in fact deliver death. Certain pitchers have a nectar laden "spoon" near the opening, while others generate motifs that are irresistible to bees and other insects, such as dark centres with radiating stripes that look like flowers. Some have backward-pointing hairs within the pitfall to prevent escape. The cobra lily has a cunning method of keeping hold of its victims. Its walls are decorated with clear patches that persuade flies to stay and beat against a window rather than flying upward to freedom. Once they have fallen into the liquid their fate is sealed, for it contains a syrup from which escape is impossible.

The slippery wax that covers the inner surface of many pitfalls has revealed its secrets. It has two layers that are loosely attached to each other.

The lower section is made of stiff foam, while the upper consists of a sheet of loosely attached and brittle plates. These break off and clog the prey's hooked feet, which then skid on the foam below. Other species keep their surface moist with a series of fine ridges that trap water or nectar and cause any insect that lands to aquaplane into the void below. They can catch their quarry only during wet weather. At other times an insect can sit happily on the rim, which perhaps lulls it into a false sense of confidence when it comes for food until, one day, it falls to its doom.

To trap a fly is just the first step. The animal must be digested and its goodness absorbed. Once again, the insect-eaters use different means to achieve a common end. In so doing they approach the habits of animals. A sundew leaf with a series of tentacles all pointing towards an item of food led Charles Darwin to "imagine that we were looking at a lowly organised animal seizing prey with its arms." Soon the whole leaf closed up to make a "temporary stomach" that smothered the prey as each hair pumped out digestive droplets, the "dew" that gave the plant its name. By a "curious sort of rolling movement," rather like that of the human intestine, the unfortunate victim was propelled towards the centre, where more hairs awaited. As it expired, the secretion changed from a mere glue to something more acid. The sticky fluid could, he found, digest living material of many kinds.

The magic liquid was pumped out from the base of each hair. There were marked changes in the internal structure of each cell as it prepared to make the digestive juice, with the accumulation of masses of purple matter after a dose of meat. The sundews gave Darwin the first hint that cells can communicate with each other, for a colour change could be tracked from one to the next as the message spread across the leaf. The agitation spread just as in animal nerves (although nerves act faster and show no visible changes).

What path did the information take as it travelled through the leaf? Various vessels traverse it, but tentacles close to and distant from such channels behaved the same way. The best that Darwin could come up with was that the motor impulse involves the passage of some kind of chemical, but what that was he did not know.

In those days all that was known about nervous transmission was that it entailed what he called an "invisible molecular change that is sent from one end of the nerve to the other," but there was no evidence of what that invisible change might be. His experiments with poisons on sundews hinted at what was to become a central truth of biology. Some, like arsenic or strychnine, were as pernicious to sundews as they were to humans, while others (cobra venom included) were not. Morphine and alcohol, with their interesting effects on our own nervous system, left the sundews unmoved—but salts of potassium and of sodium had contrasting effects. The former caused the leaf to respond, while the latter was fatal. That observation presaged the discovery that a balance between the two on either side of the cell membrane is behind the electrical activity of both plant and animal cells, nerves included.

The flytrap, with its need for a double tap before closing, must in some sense remember the first before responding to the second. Burdon Sanderson, the professor of physiology at University College London, suggested to Darwin that a chemical or electrical mechanism was involved here, too. He was the first to find that—just as in animal muscles when they contract—the voltage altered when the trap snapped shut. Now we know that a complicated long-chained sugar, when applied in minute quantities to the leaf, causes the prison walls to close. It builds up quickly after the trap is touched but is also broken down quickly. Only if the second tap arrives in time does it reach a concentration high enough to trigger a response. The memory molecule—which is what it is—activates channels in the cell membrane that transmit sodium and potassium ions, generating an electrical signal that fires off the poised cells. As in nerves and muscles, the movement of calcium ions is also involved. The ambush can also be induced with adrenaline and human neurotransmitters—molecules that themselves work by reaching a threshold—as a further hint of unexpected parallels between the plant and animal worlds.

Concerned as ever with the state of his own intestines, the great naturalist turned to Burdon Sanderson for advice about the sundew's digestive juice. Its power to break down protein had long been known. The sticky secretion of butterwort was once used to make "ropy milk," a sort of yoghurt in which milk was curdled by its digestive enzymes. Herbalists still insist that the substance works against tuberculosis, asthma, intestinal pain, and chapped udders in cows (in Holland it was popular as a hair pomade).

The two scientists established that the exudate contained a mixture of organic acids related to vinegar, together with an enzyme. When both were present—but not when just one was available—insect flesh was broken down. The sundew stomach, if it could be so called, hence showed close parallels in its workings to our own (which contains both acids and enzymes). That, Charles Darwin felt, was a "new and wonderful fact in physiology" that brought plants and animals together.

The digestive enzymes of the insect-eaters have revealed more of their secrets. Our own gastric talents are limited by comparison, for the plants cope with a diet that would give us all dyspepsia. Vincent Holt's neglected 1880 work *Why Not Eat Insects?* contained recipes for such nutritious dishes as *larves des guêpes frites au rayon* (wasp grubs fried in their nests). Some people do eat larvae—silkworms are popular in the Far East—but some of Holt's suggestions, such as *phalènes a l'hottentot* (moths in butter) and *cerfs volants à la gru gru* (stag beetles on toast) would surely be indigestible. The Vietnamese who feast on water beetles or scorpions have to throw away the tough outer shells. Their botanical counterparts can afford to be less fastidious, for their enzymes break down almost everything the prey has to offer.

The insectivores use a cocktail of chemicals, each of which digests a particular foodstuff. Sundews have an enzyme to cut up nucleic acids, while pitchers—which can hold three litres of digestive fluid—have half a dozen distinct kinds that attack proteins, nucleic acids, and other substances, together with a special protein that breaks down the insect skeleton. The Venus flytrap has equivalents (although if a leaf overeats, it can die from indigestion) as do almost all the other botanical carnivores.

For plants and people alike, digestion is followed by absorption. The insectivores have refined their abilities to soak up a meaty soup, but at a price. The leaves of most conventional plants are fairly impermeable because of the need to conserve water, but a creature that feeds through its leaves cannot safeguard itself in that way. The sundew's leaves have a generally waterproof skin, but the surface is decorated with large pores that allow its insect broth into the digestive cells. In pitchers, the whole interior covering is thin or is scattered with holes that allow water to pass. Such plants then face a dilemma, for as they suck in the liquid remains of their feasts across a porous leaf surface they are at risk of losing fluid from the same place. As a result, many are restricted to wet places.

They have to make many other compromises. First, they face a conflict between sex and food. They eat insects but are also pollinated by them. To reduce the chance of error, their flowers and their traps open at different times, or on different parts of the plant, or attract a different set of visitors. Even so, the carnivores often devour their winged Cupids by mistake (or perhaps because they are more valuable as a source of nitrogen than as a sexual aid).

What struck Darwin most about his insect-eaters—and he experimented on the underwater kinds as well as those on land—was that all built their specialised machinery by hijacking talents found in more orthodox species.

They were a wonderful example of how evolution could make do and mend. Natural selection had scavenged its raw material from whatever was available rather than waiting for something new to turn up.

Varied as those creatures are, and distinct though their tactics of trapping, digestion, and absorption might be, carnivory has always been cobbled together. All the species studied at Down House, and the many more now known, reached their present state by modifying the banal talents of their ancestors. All roots make mucus, sticky as it is, and sundews are related to tamarisks and knotweeds, whose leaves make lots of the stuff to get rid of salt or to fight off insects. Most plants are attacked by insects and some have evolved defences such as adhesive hairs or spines that can be turned to aggressive ends. The familiar blue plumbago is notorious for sticky flowers that trap its enemies and save it from attack. Many flowers close upon a pollinator before releasing it and—as a hint of how the Venus flytrap evolved its remarkable talent—plenty of plants can move their leaves or seed pods, some at great speed. The pitchers had rather little to do to develop a trap, for leaves fuse for many reasons and a variety of mutations in crops such as maize cause once-independent leaves to bond together.

Digestion, too, has echoes in more innocent parts of the botanical world. Some wild geraniums are covered in glue that can break down and absorb proteins placed upon their leaves. They may be "proto-carnivores," poised on the edge of that habit but not yet committed to it. Other parts of the digestive toolkit are also lying about, waiting to be used. Seeds and leaves secrete enzymes to protect themselves against attack. Those of certain insectivores resemble proteins that, in most species, are found only within the cell, suggesting that the leap to eating meat did not involve some novel chemical but just a talent to pump an existing one out. The sundew enzyme that cuts up nucleic acids looks rather like those secreted by all plants after damage. It too was hijacked for a more sinister end.

Most leaves can absorb some molecules, small and sometimes large, through their surfaces. Darwin himself found that even species that never eat insects, such as primulas, could transfer nitrogen-rich substances across their leaves. His observation led in time to the idea of "foliar feeding." Instead of adding fertiliser to soil, where it is washed away or rendered useless through chemical change, the hope was to spray it onto leaves, whence it would enter the plant. In alkaline places there can be plenty of iron and manganese—both needed for healthy growth—but bound into soil compounds that will not release them. For some organic gardeners the idea has become almost a cult, but it gives real benefit only when rare nutrients are missing. Some places lack zinc, or copper, or boron, all needed in minute amounts. A quick spray does a lot to help. The technique is useless for nitrogen, which is needed in larger quantities than leaves are able to soak up.

Carnivory is just a step—a case of the biter bit—in the endless battle between the insects and their vegetable prey. Every one of the tactics needed is in use somewhere else for a different reason. Evolution had only to put the package together.

The relationship between such plants and their prey shows a clear divergence of interest. Even so, conflict can shade into what seems like cooperation. Many insect-eaters depend on third parties to help them. The North American pitfall known as the Virgin Mary's socks (from its purple colour and the footwear of the pope) has no digestive enzymes of its own and depends on bacteria to do the job. A South African species that at first sight looks like a typical member of the group has sticky hairs to trap insects—but it gains its nutrition at second hand. A bug lives on the leaves and feeds on the corpses, and its excreta feed its host.

The struggle for existence between fly trapper and fly—nature red in leaf and glue—is easy to observe. Some insects, though, live not as prey for plants but, like the South African bug, in apparent harmony with them. They bear a more subtle message about how conflict drives evolution to shared ends.

Certain ants defend their hosts against attack, a talent known to the fourth-century Chinese, who put their nests into lemon trees. Once again, the tie between the two kingdoms has prompted the evolution of some remarkable organs, each emerging, like a snap-trap or a flypaper, from a distinct part of the anatomy. A celebrated passage from *The Origin* reads: "If it could be proved that any part of the structure of any one species had been formed for the exclusive good of another species, it would annihilate my theory, for such could not have been produced through natural selection." The helpful ants might appear to be such a proof, but far from annihilating Darwinian theory, they support it. On the way they hint at how insectivory began, for they pay a good part of their rent not with aggression but with nitrogen. They add a whole group of new members to the grand botanical convergence aimed at solving the fertiliser problem.

Thomas Belt was an engineer and naturalist who spent five years in charge of a gold mine in Central America. His 1874 book *A Naturalist in Nicaragua*, which Charles Darwin called "the best of all natural history journals," notes how some trees in the Acacia family had fallen into an association with certain ants, which protect them from grazing insects and mammals. As Belt saw, the balance of advantage between the two parties is finely poised.

The total mass of ants in a typical patch of Amazon jungle is four times the combined weight of its mammals, birds, reptiles, and amphibians. Certain plants have put the animals to work. Many tropical trees have hollow thorns which shelter the stinging insects, together with small bodies filled with sweet and oily material. Both partners benefit, for any creature that dares to browse on the tree is attacked and the ant gets a free meal. If its garrison is killed off with insecticides, the tree is attacked by grazers at ten times the previous rate. The visitors also prune back branches of nearby trees that shade their host, and clean up the ground around its trunk, reducing competition for food. Some even poison nearby plants by injecting formic acid into the leaves, allowing their own host to flourish on patches of cleared ground known to the locals as "devil's gardens," as they are thought to be cultivated by an evil spirit. In return the insects feed on secretions from the acacia's leaves and feed their young from what Darwin called its "wonderful food bodies." They also gain protection by laying eggs inside the hollow thorns.

More than a hundred groups of tropical trees, and forty families of ants, have entered into such relationships. It has evolved many times and, like insectivory, has enabled natural selection to adapt a diversity of parts for novel uses. The shelters may be based on thorns, hollow stems, rolled-up leaves, or special pouches made on the surface of the leaf. Once again, evolution makes do and mends as it must.

That tropical liaison gives further proof of Darwin's insistence that natural selection gives nobody a free lunch. At first sight, the bond between ants and trees is based on a shared dedication to a common end. In fact, each tries to get the most out of the arrangement while putting in the least possible effort. As they do so, they hint at how the tie between the botanical carnivores and their insect prey may have begun.

Often, the plant makes the special food only when enough ants are around to make it worthwhile. Some trees are even more parsimonious. The whistling thorn of Kenya uses ants to keep hungry giraffes at bay (it gets its name because the wind howls through the hollow thorns). It gives only shelter, and not nutriment, to its resident army. Those soldiers cheat just as much, for some get a meal from the honeydew made by scale insects that feed on sap (and that do nothing of benefit to the tree) while others have little interest in attacking herbivores. One species is even more selfish, for it castrates its host by chewing off flower buds to ensure that the plant does not waste its efforts on show but puts out new and tasty shoots instead. The tree fights back by making a chemical that keeps the ants off the flowers. Yet another insect destroys its host's food bodies to dissuade more aggressive species that might protect the tree.

The ants gain sugars, based on carbon, from their host—but their corpses and those of their prey provide precious nitrogen to the tree. The shelters have thin walls through which their excretions, or the remains of their bodies, are absorbed by the host. Some Acacias take nine-tenths of their nitrogen from insect visitors. It would not be hard to transform an arboreal nesting site into a trap that soaks up nitrogen while giving nothing back.

Acacias, like sundews, are nitrogen hunters that depend on other creatures for help. The entry of the ants makes that already diverse clan even wider than before. A further look around the botanical world shows that the tactics of acacias or Venus flytraps are trivial in comparison with the ingenuity shown by other species. Many thrive in what would otherwise be famine conditions thanks to a series of obscure but intimate relationships that provide them with the essential element; relationships not with insects (which, as animals, are relatively close to them in evolutionary terms) but with bacteria and fungi around their roots that pull the gas from the air and receive food and shelter in return. This symbiosis is central to the survival of life on earth and represents a series of convergences between minute creatures very distant from one another on the evolutionary tree.

Below the surface, nitrogen is often bound into compounds that refuse to give it up. The roots of many plants take advantage of the ability of certain bacteria to transform these compounds into a more digestible product by secreting chemicals that attract them, and soaking up their invaluable wastes. Peas, beans, and certain trees have entered into a more intimate arrangement for they are linked to specialised bacteria that enter the plant tissue itself and combine nitrogen gas in the air with hydrogen to form ammonia and other compounds that can be soaked up. Many of the insect-eaters and ant-exploiters, with their spectacular adaptations above the ground, also depend heavily on this less visible pact with a third party within their roots.

Farmers take advantage of such bacteria when they rotate their cereal crops with legumes such as clover and soybeans, for they too are nitrogen fixers. Together, such plants generate half the nitrogen used on the world's farms. Without them we would starve. Their allies make an enzyme that forces the sullen molecules of the gas found in the air into a marriage with the active hydrogen ions made as food is broken down. The reaction costs both the bacteria and its host a great deal of energy.

Before today's technical developments in biology, the bacterium involved, and the protein that does the job, looked more or less the same in each of the thousands of species that fix nitrogen. That is not true. Many unrelated plants, and even more of their minute helpers, have taken up the pastime. DNA shows that bacteria, unimpressive as they might appear, are more diverse as a group than the kingdoms of animals and plants put together. The nitrogen fixers span a good part of the spectrum of bacterial life. They are joined in their helpful habit by fungi (more closely related to us than are the bacteria) and by members of an entirely distinct kingdom of single-celled beings known as the Archaea that teem in hot springs, deep sea vents, and soil. The sea, too, is full of gas fixers, the majority scarcely known.

Most of the bacteria involved live alone for most of the time. When they come into contact with a root of the right kind, a certain sugar locks into a receptor on the root's surface. The bacteria squeeze in, and the host's cells divide to produce a nodule filled with the invader's descendants—a billion or more from a single founder cell. Both parties benefit: the plant provides fuel for the chemical work needed to drag the crucial element from the air, while the bacteria churn it out in a form that the plant can use.

The association between the two appeared, in evolutionary terms, not long ago, just after the destruction of the dinosaurs. There was, at about that time, a sudden rise in carbon dioxide and a spike in temperature, both of which favour plant growth—which meant that a sudden shortage of usable nitrogen in the soil made it worthwhile to enter into the arrangement. It has evolved again and again in distantly related families. Alder trees (but not their close relatives the beeches) have root nodules that contain bacteria better known as the producers of the antibiotic actinomycin. With their help the trees grow in starved ground such as that on dunes or mountains. Tropical ironwoods have a similar association, as do a few members of the rose and pumpkin families. Liverworts, certain ferns, and the giant rhubarb of Brazil, too—all benefit from the ability to use single-celled creatures to soak up the vital gas.

The bacteria within the roots of beans, clover, and the like have been much studied because of their economic importance. Hundreds of different kinds of single-celled helpers have been pressed into service. Some are tied to a single host—or even to a particular pea variety—while others are promiscuous. All operate under a mask of similarity, for the biochemical mechanisms involved and the molecules that signal willingness to enter into an association are diverse indeed.

Like ants on acacias or insects buzzing around a pitcher, the system shows a fine balance between cooperation and conflict. Some bacteria enter their hosts through wounds, hinting that they were once agents of infection, while a few are related to known pathogens. Others grow within a membrane that protects them from attack, or make poisons that suppress a host's ability to fight back. The plants have stayed suspicious of their partners. Now and again a cheat gets in—a bacterium that produces little usable nitrogen but demands free food and shelter. The host cuts off supplies, the nodule withers, and the fraudster starves.

Nature's market in nitrogen turns over billions of tons each year, which passes from air to soil, from land to water, and from plants to animals and back again in an endless cycle. As in all markets, the accounts of profit and loss are carefully checked. The struggle for the element is as unforgiving as that for water, air, or sex, but only now and again are its dealings exposed in all their brutishness. Plants that eat animals are just one instance among many to show how competitive that business must be and how the most improbable expedients are pressed into service to squeeze the most out of what is on offer.

Now the global trade in nitrogen has been revolutionised by man. Farmers pour fertiliser onto the soil. They buy it from factories that each year generate a hundred million tons of the stuff from oil or by extracting gas from the air. The reaction is carried out with the help of catalysts in boilers held at high temperature and extreme pressure. Without that technology, invented only a century ago, the world would starve. The industry is profligate in its use of power, most of it gained from burning the remains of ancient life. Cars, chimneys, and aircraft also pump nitrogen salts into the air. All this means that much more of the material is available in usable form than in Victorian days. The amount has doubled in the past century.

Adding fertiliser to fields certainly increases yield, but it also changes the economics of the plants' arrangement with the living sources of raw material. First, it alters the balance of profit and loss. Fertilised crops need less help from their single-celled assistants in the soil and squeeze them out at once. As a result the amount of the element taken from the air by those useful creatures goes down, so that the overall gain from the added nitrogen is much less than it might otherwise be (which is a particular problem in the starved earth of Africa).

Often, the excess is washed to where it is not wanted, and more is added by rain, itself full of nitrogen salts emitted by exhausts and chimneys. The insectivores and ant-shelterers now have a cheaper source of the missing element than they did before. The rain-fed bogs of New England were once full of pitcher plants that flourished as they sucked up nutriment from their prey. Their competitors could not manage in such starved places. The marshes are now being enriched. In those hardest hit—near cities or fertilised fields—the insect-eaters have begun to give up their carnivorous habits in favour of a conventional existence. Other species are moving in and are driving the pitchers and Venus flytraps to extinction. The same has happened to acacia trees that grow close to farms. In Europe the sundew faces the same problem so that the insectivores share a fate in death as they did in life.

Carnivory, which began with shortage, may perish with excess; and insects, at least, can breathe a sigh of relief. For biologists, too, the way habits of nitrogen-fixing bacteria and fungi, of ants and thorn trees and of the Venus flytrap and the sundew, bring good news, for they are a powerful proof that under natural selection parallel lines often converge.

CHAPTER 3

Shock and Awe

MANY AMERICAN COMMENTATORS HAVE GLOATED THAT ZACARIAS Moussaoui, the Frenchman accused of involvement in the September 11 attacks, will certainly go mad as a result of his solitary confinement in the Colorado maximum security prison where he will spend the rest of his life. As the judge who passed sentence said: "You will never get a chance to speak again . . . and will die with a whimper."

Men do fall into insanity in such places, but much as vengeful right-wingers might celebrate such mental decay, some among them would be dismayed to learn that Moussaoui will lose his mind for Darwinian reasons. Guy the Gorilla, star of London Zoo in the 1950s, was admired for his solemn disposition. In fact the animal was deeply depressed, kept as he was for years alone in a small cage. *Homo sapiens* is a social primate, descended, like gorillas or chimpanzees, from an ancestor with the same habits. Had our forefathers been more solitary beasts like the orangutan (which spends much of the year alone), the worst of all punishments would not be solitary confinement but an endless dinner party. The constant exchange of subtle emotional cues around the table would drive those present to their wits' end.

Scientists are often asked to explain what makes men different from chimpanzees or orangs, but in some ways that is scarcely an issue for science. Such questions deal not with the body or the brain but with the mind, a topic that many biologists consider to be outside their expertise. Even so, as science compares man's behaviour with that of his relatives, it finds that biology says something about how humans became what they are. *Homo sapiens* is, says all the evidence, a creature that craves company. To satisfy that yearning, men and women spend much of their time in silent and sometimes subliminal conversation. Those who for one reason or another cannot join in pay a high price.

To Sartre, hell was other people. Rousseau, too, saw man as in decline from a pure and animal state and modern society as a corruption of what the world should be. "Savage man, left by Nature to bare instinct alone . . . will begin with purely animal functions. . . . His desires do not exceed his physical needs: the only goods he knows in the Universe are food, a female, and rest." The true life was near-solitude, on a remote island best of all, with a bare minimum of interaction with anyone else. The philosopher's ideas were romantic but wrong. Members of all primate communities, human or otherwise, must negotiate to maintain peace, have sex, and reap the benefits of cooperation. They use signals both self-evident and subtle to test the mental state of their fellows and to advertise their own (and even the solitary orang hoots now and again to impress its neighbours). Civilisation is based on the ability to respond to other people's sentiments.

In 1879, at the Derby, Darwin's cousin Francis Galton found he could assess "the average tint of the complexion of the British upper classes" as he observed the crowd through his opera glass. Then the race started, and in a letter to *Nature* entitled "The Average Flush of Excitement" he noted that that average complexion became "suffused with a strong pink tint, just as though a sun-set glow had fallen upon it." A shared hue was a statement of a common passion, and Galton could work out what it was even when he could not distinguish one person from the next. In the same way, someone exposed to an image of a group of individuals bearing a range of expressions from happy to miserable can sense their general state of mind far faster than he could by scanning each visage. Our brain, it seems, has a filter that picks up not just how many individuals are in a crowd, but how they feel. The ability has its downside. It means that mass hysteria can spread through society as shared feelings feed on themselves; as Charles Mackay put it in his 1841 book *Extraordinary Popular Delusions and the Madness of Crowds*, an account of the South Sea Bubble and other fantasies, men "go mad in herds, while they only recover their senses slowly, and one by one."

In 1872, in *The Expression of the Emotions in Man and Other Animals*, Charles Darwin discussed the role of signals in the herds, packs, flocks, schools, towns, and cities in which social animals live. The book was a first attempt to understand sentiment in scientific terms. Darwin was interested in how mental actions are manifest in the face and body, and he realised how much the displays of inner feeling made by men and women resemble those of animals. The book discusses instinct, learning, and reflexes in creatures as different as moths and apes. Its author knew that elephants wept and hippopotami sweated with pain. When he heard a cow grind her jaws in agony he was reminded of the gnashing of teeth in hell. He saw that loneliness, fear, or anger and their outer signs have all—like limbs or eyes—evolved. Kick a dog and it crouches and turns down the corners of its mouth; torture an al-Qaeda suspect and he does the same. *The Expression of the Emotions* makes a powerful case for the shared mental descent of humans, primates, dogs, and more.

Our own sentiments have long been compared to those of other creatures. The seventeenth century painter Charles Le Brun (whom Darwin referred to as a pioneer in the study of emotions) urged those who wished to portray their subject's mood to first scrutinise beasts. A few hours with swine, lascivious, gluttonous, and lazy as they were, would help one depict the inner life of a debauchee. Darwin's friend George Romanes went further, setting out a scale from one to fifty. Worms and insects came in at eighteen, as they could feel surprise and fear, while dogs and apes were equal at twenty-eight, as each had "indefinite morality along with the capacity to experience shame, remorse, deceit and the ludicrous." The scale reserved points twenty-nine to fifty for men and women of greater or lesser virtue.

Psychology is still marked by such ideas. *Expression*'s central theme was, as ever, a world in which all of life's attributes, from anatomy to anguish, emerge from shared descent. Science still uses that logic on elephants, cows, apes, fruit flies, and bacteria in its attempts to build a shared narrative of inner feelings. Those who transmit their sentiments expect a response from those who receive them. That two-way commerce involves a need to acknowledge, to copy, and to respond. People gasp in sympathy at a sad tale, gaze at where another person's eyes are directed, or avoid food that someone else has rejected. Such reflections of another's mental state are part of what makes us human.

Darwin, a practical man, had little interest in philosophy. Even so, he realised that the biology of the mind was harder to interpret than that of the body. He wrestled with the issue much as modern psychologists try to come to grips with their own sometimes murky ideas. Can our thought be explained as "direct action of the excited nervous system on the body, independently of the will," and if so, what (if anything) does that mean? Shakespeare writes of Cardinal Wolsey, "Some strange commotion / Is in his brain; he bites his lip and starts; / Stops on a sudden, looks upon the ground. / Then lays his finger on his temple . . ." Such behaviour, Darwin felt, came from the "undirected overflow of nerve-force"—but is that phrase just an attempt to avoid deeper and less tractable questions? The great man's task was made harder by his quarrel with the antievolutionist Charles Bell, author of the standard text on facial anatomy. Bell was convinced, quite wrongly, that humans had unique muscles divinely designed to express morality, spirituality, or shame—a notion not much help to someone anxious to understand the smile or the blush, but an example of the preconceived truths that still plague attempts to understand the human mind.

After a long stumble through the Freudian fog, the study of the mental universe has once again become a science (although claims to have found the neural foundations of society do not yet deserve that status). Physicists and chemists now busy themselves with questions once raised only by intellectuals. In institutes of psychiatry, neurology, and zoology, cats, mice, and dogs are used to dissect human habits. Even bacteria behave in what might appear to be a rational fashion when they settle down close to a source of food, or join hands with their colleagues to form a sticky film over teeth or wounds. Certain fruit fly genes lead to homosexual behaviour and others to loss of memory (which may help in the study of Alzheimer's disease). Experiments on the brains of mice and monkeys, once done with a scalpel, are now carried out with machines of fantastic complexity. Such equipment is also used on people with brains damaged by strokes or accidents, while drugs help understand the mental universe of the normal, the reckless, and the insane. Plenty of the questions raised in *Expression of the Emotions* have a modern air. Many remain unanswered.

Emotions is in some ways a less satisfactory work than Darwin's plant, barnacle, or earthworm books, and an unusual note of apology creeps in: "Our present subject is very obscure . . . and it always is advisable to perceive clearly our ignorance" (and in that the author was franker than some of his successors). He soon found that even what seemed simple—the objective description of the facial expression of a man or a dog, for example—was hard, while to represent the sentiments behind it was even harder. That problem, in spite of the wonders of electronics, still baffles neuroscientists. Darwin was suspicious of phrenology—the notion that particular segments of the brain are associated with obstinacy, pride, guile, or other traits (an admirer claimed that the great naturalist's own head had "the bump of reverence developed enough for ten priests")—but he struggled with the question of just where felt experiences might be seated.

He looked first at the animals and children of his own household. As a kindhearted parent, he was careful not to disturb them too much (although his book contains images of frightened babies that would see him accused of cruelty today). He did not hesitate to play the animal himself: his son Francis remembered that his father's body was very hairy, and that the great man would growl like a bear when his children put their hands inside his shirt.

Even in play the *Beagle*'s naturalist was serious, and he soon identified some general rules of human and animal behaviour. Intimations of happiness or grief, of welcome and rejection, and of other opposing sentiments often came as mirror images. Thus, a frown is the opposite of a smile and a look of surprise the converse of a greeting. Some gestures emerged from movements that once had functions of their own, so that begging with open hands is related to the posture taken when holding food. In the same way, a person who rebuffs an advance will close his eyes and look away, as if from an unpalatable meal. Animals followed the same rules, and the paterfamilias of Down saw almost the same downcast looks in his household pets as those adopted by his infant son.

From such simple observations emerged the science of comparative psychology. It began with dogs.

Pets gain their status because they seem (to their owners at least) to be almost human. Darwin was no exception and kept a dog—Sappho by name—even when he was a student. He saw no problem in describing canine sentiments in the same terms as our own. His pet when in "a humble and affectionate frame of mind" acted quite unlike an animal in hostile mood with its bristling hair and stiff gait. The "principle of antithesis" was hard at work and opposed sets of muscles were set into action to express contrasting emotions. The "piteous, hopeless dejection" of his favourite hound, when it discovered that it was not about to go for a walk but instead was to sit in on an experiment in the greenhouse, was manifest in a "hothouse face," the "head drooping much, the whole body sinking a little and remaining motionless; the ears and tail falling suddenly down, the tail by no means wagged." That contrasted with its expression when happy and excited, with head raised, ears erect, and tail aloft.

As well as such individual shifts of mood, the proud pet owner noted marked differences in personality among breeds. Descent with modification could, he realised, apply to minds as much as to bodies. Certain kinds, like the terrier, grinned when pleased while others did not. Spitz dogs—huskies, elkhounds, and the like—barked, while the greyhound was silent. The canine universe encompasses a wide range of talents. Some varieties herd sheep and cattle (and, in the case of the Portuguese water dog, chivvy fish) while others guard, hunt, guide, or annoy the general public. The various breeds when taken together show a wider range of behaviour than that found among all wild canines around the world. Many of the differences are innate, and *The Origin* tells of a cross with a greyhound which gave a family of shepherd dogs a tendency to hunt hares. So impressed was its author with their divergence in habits that he suggested that some of the household types had descended from distinct wild ancestors (and there he was wrong).

His favourite pet is now back at the centre of the emotional stage. The world has four hundred million dogs, and the wonders of science have transformed the creatures into a gigantic experiment on the biology of sentiment. Even in the brief period since modern breeds began to emerge in Victorian times, dogs have undergone large—and inherited—changes in temperament.

Men long ago began to use the animals in the hunt. They soon learned to choose those with special abilities—to track, to run, to squat into a "point" position when prey is spotted, or to bite and tear or recover corpses—as parents for the next generation. Such remnants of the chase live on in the behaviour of bloodhounds, pointers, setters, retrievers, and bull terriers. Herding dogs such as Border collies stalk a sheep and do not bite it, but those used to control larger animals—like the corgis once used with cattle—go further through the sequence and snap at their charges. Pit bulls complete the job and will hold a bull by the nose (and as a side effect sometimes kill their owners). Guard dogs such as Pyrenean mountain dogs, whose job is to frighten off predators, have given up the hunt altogether. They play like huge puppies and show little interest in their herds, but their conduct is odd enough to persuade wolves to stay away. Such differences emerge from inherited variations in behaviour within the common ancestors of each breed, from new genetic errors as the generations succeed each other, and from the accumulation of change by human choice.

One way to assess a dog's personality is to startle it by introducing a stranger. Does the animal play with the visitor, back away, sniff him, or chase him out of the room? Does a sudden noise anger the beast, terrify it, or leave it unmoved? Other tests include the ability to sit still, to cope with solitude without whining or panic, to run through a maze, and to find hidden food.

Cocker spaniels are calm and obey orders, while basenjis are nervous and almost impossible to train. Crosses between the two suggest that the difference in their nature is inborn, for the offspring have a range of talents, intermediate between each parent. A survey of ten thousand German shepherds and rottweilers in Sweden showed, within each type, a shared inheritance of excitability, tail-wagging, and the tendency to bark, while aggression appears to be under separate control. In an echo of *The Expression*'s principle that antithetical emotions are expressed as mirror images, variation in all those capacities turns only on how shy or bold a particular breed might be.

As dog fanciers' tastes were further refined, more and more specialised varieties emerged. Some developed habits that perturbed their owners. Mating like with like had exposed rare and once-hidden genes, many of which had undesirable effects on personality. A few have parallels in the mental lives of men and women. In an echo of human obsessive-compulsive disorder, some bull terriers chase their own tails for hours until they collapse, while springer spaniels may savage their masters in a sudden attack of uncontrolled rage. Certain families of basset hounds suffer from a delusion reminiscent of paranoid schizophrenia and cower at the slightest noise. Some Dobermans, in contrast, fall into a heavy slumber after an unexpected snack. They have narcolepsy, a distressing and sometimes dangerous condition also found in people—and the dogs respond well to the drugs used to treat human patients.

The double helix reveals why breeds diverge so much in personality. The first complete sequence came from a boxer. With about twenty thousand genes the animal had several thousand fewer units of inheritance than humans have (and fewer even than the mouse). One hope was to find canine matches to our own disorders, and some have already emerged. The sleep problem in Dobermans involves damage to a certain receptor protein on the surface of brain cells—and the human equivalent reflects a fault in the same gene. No doubt our fireside companions will help track down many more of the inherited errors behind our own mental illnesses, as they already have for conditions such as blindness. Darwin would be proud.

Dogs are anomalous animals, for their habits have been so altered by human effort that their mental universe is far from that of a wild creature. Charles Darwin soon moved on in his search for the roots of human emotion. He spent many hours at London Zoo and had particular fun with the anatomy of amusement: "Young Orangs, when tickled, likewise grin and make a chuckling sound . . . as soon as their laughter ceases, an expression may be detected passing over their faces, which, as Mr. Wallace remarked to me, may be called a smile. . . . I tickled the nose of a chimpanzee with a straw, and as it crumpled up its face, slight vertical furrows appeared between the eyebrows. I have never seen a frown on the forehead of the orang." He was much taken by the attempts of a monkey to court its own image in a mirror and by the antics of Jenny the Orang-Utan, who when teased by being shown an apple through the bars "threw herself on her back, kicked and cried, precisely like a naughty child."

Primates, like people, reveal their feelings on their faces. Someone who has never seen a macaque can at once identify its mood when shown photographs of the animal expressing emotions such as sadness, happiness, or rage. Many chimpanzee expressions have been named. They include the closed-mouth smile, its bared-teeth equivalent (which descends from an ancestor shared with our own smile), the bozo smile (a rather inane grin), and the play face (a relative of human laughter), together with subtler statements of mood such as the stretched pout-whimper. Bonobos have an amused expression and make a noise that is uncannily like a guffaw (a German expert has also identified an *Orgasmusgesicht*, or "orgasm countenance," in that species, although its existence in humans remains to be demonstrated). Gorillas are more impassive; they grin and make bozo faces but otherwise keep their feelings to themselves unless they are enraged.

Apes and monkeys can also interpret their fellows' moods. Electronic chimp avatars can be manipulated to simulate pout-whimpers and other emotions. When real animals are presented with their artificial comrades, they pick out different expressions at once (screaming faces best of all). They also show some insight into another's feelings. If one animal sees another grimace in fear when it hears a buzzer it has learned to associate with an electrical shock, the observer will flinch at the sound even though it has never itself experienced the shock.

Monkeys and apes reflect their moods in their postures as well as their expressions (gorillas really do slap their chests in rage). Humans, orangs, chimps, and gorillas share the Italianate habit of waving their hands. Bonobos flap their wrists in irritation, point at themselves when they need a hug, and hold out their palms when food is on offer. In a further nod at our common heritage, they prefer to signal with the right hand.

People are even better at reading each others' sentiments. We are so attuned to human features that we often see them when they are not there (which explains the much-commented upon apelike countenance in NASA's early pictures of the mountains on Mars). Two crumpled newspapers look much the same although their shapes are quite different, while two faces are seen as quite distinct although their shapes are almost the same. A simple bar code, the position of six stripes of dark and light—hair, forehead, eyebrows, nose, lips, and chin—conveys almost all the data. Most of us can recognize thousands of individuals and sense dozens of emotional states. Faces are important even to infants. Darwin noted that early in life his children spent long periods gazing at their mother. Now we know that a baby responds to a human countenance—even in a photograph—within minutes of birth.

Men, like apes, speak with their faces and in much the same language. Angry people and angry gorillas bare their teeth, and a frightened chimpanzee looks rather like a frightened person. For humans (as for apes) some expressions are ambiguous. Men and apes bare their teeth when amused but do the same when terrified. *Emotions* has a picture of a Sulawesi macaque grinning in pleasure as it is stroked—but in other macaques the same gesture marks submission to a threat. Not all our grimaces are shared with our relatives. Apes never signal disgust, and their noses, which are more sensitive than ours, remain unwrinkled even to a repulsive smell. A wide-open mouth is a threat in many primates but conveys only mild surprise for humans, and while elephants weep our closest kin do not.

Even so, our own faces are more eloquent than those of any other primate. Many pages of *The Expression of the Emotions* are devoted to the way they reflect their owners' inner state. Some seem quaint nowadays: "The breach of the laws of etiquette, that is, any impoliteness or *gaucherie*, any impropriety, or an inappropriate remark, though quite accidental, will cause the most intense blushing of which a man is capable. Even the recollection of such an act, after an interval of many years, will make the whole body to tingle. So strong, also, is the power of sympathy that a sensitive person, as a lady has assured me, will sometimes blush at a flagrant breach of etiquette by a perfect stranger, though the act may in no way concern her." In the interests of science, modesty gave way to the search for truth: "Moreau gives a detailed account of the blushing of a Madagascar negress-slave when forced by her brutal master to exhibit her naked bosom," and the sexual nature of that expression means that "Circassian women who are capable of blushing, invariably fetch a higher price in the seraglio of the Sultan." Mark Twain, an ardent evolutionist, put it well: "Man is the Animal that Blushes. He is the only one that does it—or has occasion to."

Darwin was keen to discover whether signals such as the blush meant the same thing in every human culture, or whether, like skin colour, they changed from place to place. Rejecting the popular notion that different races had evolved from higher or lower primates and that their mental lives and expressions of mood reflected this, he accumulated a mass of anecdotes that made the case for the universal nature of facial cues. In addition to William Ewart Gladstone, who commented on statements about the Greek visage found in Homer, his correspondents included "Captain Speedy who long resided with the Abyssinians; Mr Bridges, a catechist residing with the Fuegians and Mr Archibald O. Lang of Coranderik, Victoria, a teacher at a school where aborigines, old and young, are collected from all parts of the colony." One letter told of a Bengali boy with "a thoroughly canine snarl." Its recipient fired off a series of questions to those servants of the Queen, sometimes to ludicrous effect ("Mr B.F. Hartshorne . . . states in the most positive manner that the Weddas of Ceylon never laugh. Every conceivable incitive to laughter was used in vain. When asked whether they ever laughed, they replied: 'No, what is there to laugh at?'").

The Weddas notwithstanding, Darwin became certain that such signs were more or less universal across the globe: "The young and the old of widely different races, both with man and animals, express the same state of mind by the same movements." Hard as it may be to believe, that observation was forgotten and for many years anthropologists assumed that expressions were determined by culture (although nobody found a place where people laughed in pain or screamed in greeting). Looks of anger, disgust, contempt, fear, joy, sadness, and surprise—all are universal (one tribe in New Guinea cannot separate expressions of fear from those of surprise—but for them any intruder is a threat). People from different societies do find it harder to identify each others' guilty or shamefaced looks than they do a smile or an expression of terror, suggesting that such subtle statements of mood may be partly learned. Even smiles are equivocal, for the beam, grin, smirk, snigger, simper, and leer each convey a different message, while people who smile too often come across as nervous rather than contented. Darwin, too, noticed some ambiguities when asking observers to interpret mood from photographs. The expression of a man almost in tears was recognised by some as a "cunning leer," a "jocund" frame of mind, or even as someone "looking at a distant object."

Once he had established that most such signs were common to all mankind, Darwin set out to describe them in more detail. Measurement, he knew, is the first step in science (a lesson much ignored on the wilder shores of psychology), and he tried hard to give impartial descriptions of human features ("The contraction of this muscle draws downwards and outwards the corners of the mouth, including the outer part of the upper lip . . . the commissure or line of junction of the two lips forms a curved line with the concavity downwards and the lips themselves are generally somewhat protruded").

In today's world of fraud, terrorism, and identity cards, such attempts to put facts onto faces have become an industry. Remarkable claims are made about the ability to identify individuals and to sense their states of mind. Some enthusiasts recognize thirty manifestations of anger and eight of sadness, with additional criteria based on how the subject holds his head. Under George W. Bush the Department of Homeland Security spent millions on machines that were supposed to detect when a terrorist attack was planned by examining the suspect's countenance. Nobody denies that the expression of a Scotsman with a grievance can easily be distinguished from a ray of sunshine, but such claims go too far.

The face says a lot about how we feel, and the body adds information to the stream of emotional cues. A man with raised fists is not about to make a visitor welcome. Psychologists tend, for practical reasons, to use pictures of faces alone. That can be a mistake. An image of a person with a disgusted expression, taken from a catalogue of facial poses, is interpreted as just that when superimposed onto a body holding a pair of dirty underpants—but as anger when added to a fist-waving torso, or triumph when stuck onto the beefy frame of a bodybuilder. The same photograph shown against the background of a cemetery is interpreted differently than when seen against a neutral surface. An assumption of simplicity can confuse results taken from complicated machines.

The face is a real mirror to the soul. Even a brief glimpse reveals the presence of another individual, identifies who it might be, and gives a strong hint as to what its bearer will do next. Most westerners interpret a set of features with a quick triangular scan of both eyes and the mouth, each of which says a lot about identity and state of mind—but Chinese concentrate their attentions instead in a fixed look at the nose, picking up the general expression of the whole visage. Electronic scans show that when someone flashes into view, the brain first notes his or her presence, then identifies who it might be, and last of all tests the person's mood: this is a face, it belongs to Fred, Fred is furious. It processes a portrait twice as fast as it does a picture of other objects. A certain part lights up about a tenth of a second after a face is first seen, notices who it is about a fifteenth of a second later, and takes even longer to interpret what humour the visitor might be in.

Some expressions are easier to identify than others. The smile is coded deep within the skull and everyone has an inborn ability to assume it. As Darwin noted, babies born blind smile without difficulty (and blind athletes raise their arms in the air in a chimplike gesture of triumph when they win). Children find it easier to pick out expressions of good cheer than they do those of fear or disgust; women smile at strangers more than do men, while men are worse at working out mood from a slight movement of the lips. A lopsided grin to the right is seen by most of us as more joyful than its equivalent on the left. Even sheep, when given the choice of a smirking or a sombre shepherd from whom to take food, prefer the cheery one. We smile or raise our arms not to reassure ourselves that we are happy or proud but to tell others how we feel. Context is all: when Chelsea football club scores, fans respond with roars of triumph rather than smiles of delight, but gold medal winners as they stand on the Olympic podium have wider grins than those who have gained bronze.

Signs of delight or terror seem simple enough, but there are real differences in the ability to decode them. I have a talent that illustrates that fact: I can waggle my eyebrows. It began in school when I was rebuked for glowering. I then tried dumb insolence with a one-brow grimace rather than the full two-brow scowl, and in time it became easy to alternate. I sometimes amuse small children with the trick—and usually they smile back. Unfortunately, an occasional infant screams instead. The signal is clear but the response uncertain.

Both steps can go wrong. Some people cannot tell individuals apart by their faces and use clues from voice or clothes instead. In one instance, a litigant wandered into court and discussed his case with a barrister—not his own but his opponent's. The context was right: a lawyer, with a gown, in a courtroom. The face alone did not fit. Needless to say, he lost. Face-blindness may be caused by a stroke, but one form runs in families with perhaps just a single gene involved.

Other unfortunates lose the ability to broadcast their emotions. For some reason—injury, infection, cancer, or brain hemorrhage—the facial nerve no longer works and the patients cannot express their feelings in their features. They find it hard to assume looks of happiness, fear, or surprise; and their wives, husbands, and friends notice the problem. The condition might seem trivial, but in fact causes real distress (sometimes even suicide), particularly when an attempt to smile emerges as a grimace or a leer because the eyebrows—usually lifted at a happy moment—refuse to obey instructions. Some people have their brows surgically moved upwards, which gives them a permanent look of surprise, while others grow a long fringe that hides the offending forehead. The willingness to take such steps shows how important signals of mood are as a passport to society.

The Expression of the Emotions marked the first real attempt by science to infer the action of the mind from its external signs. Scientists now study the activity of brain cells rather than of facial muscles as they try to understand our inner feelings. The use of electricity—and of the sophisticated devices that depend on it—in psychology has become a science of its own. It was first expounded in that great work.

The ancient Greeks used electric fish to treat headaches, but for many years the galvanic fluid was no more than an entertainment: an entire community of monks was once connected by a mile-long iron wire and made to jump for the amusement of the King of France (castrati were tested to see if they acted as insulators, but they did not). *Emotions* contains several pictures of faces stimulated by shocks to give expressions that resemble the natural look of horror, rage, and the like. They came from the French physician Guillaume-Benjamin-Amand Duchenne de Boulogne. Duchenne is best remembered for the muscle-wasting disease named after him, but he also studied the expression of what he called the "passions" by touching electrodes to different parts of a countenance. He was the first to notice that a genuine smile involved raising the eyebrows, and his machine could easily activate those "sweet muscles of the soul" to simulate a happy beam. He even made the visage of a decapitated criminal assume a simulacrum of pleasure with a probe upon its cheek. Duchenne chose as his main subject an aged man of feeble intellect, for he "wanted to prove that, despite defects of shape and lack of plastic beauty, every human face can become spiritually beautiful through the accurate rendering of emotions." His pictures played an important part in Darwin's attempts to give an objective account of signs of pleasure or pain.

The machines have marched on. Where Duchenne used a battery, a metal rod, and a plate camera, scientists in search of the springs of sentiment now turn to the electro-encephalogram (EEG), positron emission tomography (PET), or functional magnetic resonance imaging (fMRI). Tiny electrodes activate single nerve cells, while the EEG and its relative the magnetoencephalograph respond to electrical activity within the skull. PET scanners use a sugar marked with a radioactive label, which is taken up by active parts of the brain, and detect its decay products. The fMRI machine picks up tiny changes of blood flow through the grey matter from a shift in the magnetic properties of haemoglobin as it gains or loses oxygen.

Marvellous as such techniques may be, they run into many of the problems that plagued the author of The *Expression of the Emotions.* He had found it hard to decide just where the jaw ends and the cheek begins, or to identify the precise arrangement of facial muscles. Today's arguments about the boundaries between areas of the brain as defined by scans—however confidently coloured and labeled—echo his uncertainties about the anatomy of the human countenance. Some claim that single emotions can be mapped to one or other part of the brain. Others see that organ—as he saw the face—as a connected structure, with most sections contributing to most of its functions. Any attempt to pinpoint single centres of anger, joy, or despair is a mistake.

Another problem for both the nineteenth and twenty-first centuries comes from the need to describe broad sentiments in narrow terms. Darwin was happy to describe his dog's "humble and affectionate frame" of mind—but how is it possible to put figures on humility or affection? Objective fact soon slides into interpretation, and *The Expression* was itself not immune from that temptation. Its photographic plates are not originals but engravings, some touched up to make a point. A mad lady with alarming hair was given a furrowed brow by the engraver, and a screaming infant was made to look even more miserable (the picture sold hundreds of thousands of copies to a gullible public). The lurid images of centres for pain, passion, and pleasure that decorate the scientific literature and leak into the press are also in a sense fakes. Digital information is processed in a sophisticated (sometimes subjective) way to make a picture that may be rather more than the sum of its parts.

A final difficulty for both the Victorians and their descendants was to find subjects willing to display their emotions to the world. Duchenne set up a theatre in which the public could be delighted by actors galvanically activated to produce an air of grief or delight. Many of *The Expression*'s illustrations also depict members of that profession. One bearded thespian looks remarkably implausible as he strikes his attitudes. Actors still play an important part in neuroscience. Their photographs are taken as they simulate a mood and are then shown to subjects whose brains are scanned to see which bits light up. Many of the images look just as posed as those of Darwin's theatrical friend. The artistes' profession depends on overdoing the job, often to a bizarre degree. People shown pictures of frightened or unhappy people taken from real life have far less of a nervous response than they do to those who simulate a mood as they strut and fret upon the laboratory floor. Most of us find it harder to interpret the sentiments (apart from laughter) in muted clips of film stars like Dustin Hoffman or Meryl Streep than the simulated joy or terror of a ham actor—and yet the hams supply the raw material for experiments of huge technical sophistication and expense.

Various claims have been made that particular parts of the brain respond to the sight of a happy or miserable countenance, or that they prepare the nervous system to beam back or look sympathetic in return. Because light comedy is a subtler entertainment than Greek tragedy, many scientists who study the great theatre of emotion focus not on mild signs of contentment or sadness but on expressions of horror and dread.

A blank stare is a signal of terror, and Shakespeare knew as much. A furious Othello says to his supposedly unfaithful (and frightened) wife Desdemona before he kills her: "Let me see your eyes." The whites of our eyes are larger than those of any other primate, and we take more notice of them. In a chimpanzee the mouth is a far more important conveyer of emotion. We process eyes quicker than we do any other feature, and fearful, widely stretched eyes even faster—and women do the job better than men. One woman could not recognize a picture of a terrified individual because she did not look at the eyes. When instructed to do so she at once understood the subject's frame of mind.

The brain's main activity in response to a frightened look takes place in a pair of structures called the amygdalae. These are almond-shaped groups of nerve cells deep within the temporal lobes, embedded into what are sometimes seen as the brain's most primitive parts. Each is connected to other brain centres, to the hypothalamus—that hormonal bridge between the nervous system and the bloodstream—to nerves that feed from pain receptors and from the eyes, and, in primates more than other mammals, to nerves to and from the face.

Animals in which the amygdalae have been damaged find it hard to pass the classic test in which fear of an electric shock becomes associated with the sound of a bell. Experiments on monkeys in which those structures were cut out showed that they also lost their ability to recognize familiar objects and shed both their nervousness about human company and their maternal affection for their own offspring. Human patients with damaged amygdalae have similar problems when faced with emotionally draining tasks.

The amygdalae are busiest when a frightened gaze is directed straight at its target—which fits Darwin's idea that a countenance stricken by terror is an immediate signal of danger. A few people have such severe brain damage that they perceive themselves as blind—but show them a scared person and the amygdala lights up. We are slower to notice the racial origin of an angry face than we are a happy one, showing that fear has priority over familiarity. In the United States, images of black people shown to whites stir up more activity than do those of individuals of their own skin colour.

The case for the amygdala as the fear-detecting organ seems persuasive but, as usual when it comes to the contents of the skull, real life is not simple. Other parts of the brain are also involved in the response to a terrified countenance. The amygdala lights up in response to a whole set of features rather than just to the eyes, and does so to some degree whether or not the subject shows signs of alarm. Its main role may be to notice new events rather than to make a specific response to a particular emotion. It is also involved in memory. People recall where they were on September 11, 2001, with the help of their amygdalae, but those in whom the structure is damaged remember the disaster no better than they remember what they had for breakfast.

The structure helps to process a nerve transmitter called serotonin (which is also involved in temperature control, sleep, hunger, lust, response to injury, liver repair, and more). Many antidepressants work by changing the way serotonin is broken down or taken into cells. Differences in the ability to respond to or make the substance might be behind individual responses to fear. Some people are terrified by the simplest problems of society. Darwin writes of a dinner party given for a man who, in thanking his hosts, "did not utter a single word; but he acted as if he were speaking with much emphasis. His friends, perceiving how the case stood, loudly applauded the imaginary bursts of eloquence, whenever his gestures indicated a pause, and the man never discovered that he had remained the whole time completely silent." The unfortunate fellow could now be comforted with the information that he may have a more active amygdala than normal and that his nervousness might be treated with drugs.

Inborn errors in the ability to synthesize serotonin make some people sad, angry, or suicidal. A gene whose product helps remove the chemical from the junctions between nerve cells comes in two common forms, one better at the job than the other. The less active type is more frequent among people who are unduly anxious, neurotic, or depressed—and its bearers are less able than their fellows to decode expressions of fear or sadness. The orangutan, the most solitary of our primate kin, has an even feebler version of the gene (although whether that has anything to do with its lonely life and presumed dislike of dinner parties remains to be proved).

People with severe depression often find it hard to sense the emotions of others. Drugs that affect serotonin can help the illness—and their most immediate effect, within hours of the first pill, is to improve a patient's ability to interpret their fellows' feelings by their faces. That simple talent helps turn the key that restores them to society.

Nowhere is the importance of signals more apparent than in children. When they are very young, their insights are limited and their minds self-centred; but soon they begin to understand and to respond to the moods of those around them. Darwin wrote a *Biographical Sketch of an Infant*, an account of child development based on his son William: "When 110 days old he was exceedingly amused by a pinafore being thrown over his face and then suddenly withdrawn; and so he was when I suddenly uncovered my own face and approached his. He then uttered a little noise which was an incipient laugh." William "did not spontaneously exhibit affection by overt acts until a little above a year old, namely, by kissing several times his nurse who had been absent for a short time." By then he could tell people apart (some of whom pleased him more than others) and could copy movements. By the age of eighteen months, most children can separate false movements of anger or upset made in play from real ones, and by five they send and receive information well enough to allow them to live in groups, to learn, and, in time, to join society. A sense of self and a sense of other are closely related, for the younger a child recognizes a picture of itself, the better it interacts with its fellows when it grows up.

William and his brothers and sisters were lucky to be raised by a kindly mother and father in an affectionate household. Many children are less fortunate. An infant brought up in isolation or by cruel parents may never adjust to the world around it and may feel isolated for its entire life. The fit between childhood abuse and adult depression is well established, and children taken into care because of poor parenting are at above-average risk of later problems.

A few unfortunates suffer for the opposite reason: what condemns them to loneliness or despair is not neglect by those who should provide the crucial emotional messages, but their own inability to receive or to interpret them. Such children are often diagnosed as autistic. The most severely affected live in isolation and unhappiness, for they cannot make or understand the cues needed to find a place among their peers. Their plight shows how central to society is the ability to express and understand another's feelings.

People with autism are now treated with sympathy and concern, but once they were regarded almost as animals. To those curious about where the essence of humanity might come from, they were useful raw material for speculation. Rousseau wondered whether a youth brought up "wild, untamable and free" would be free from the corruption faced by those who undergo a normal education. He pondered an "impossible experiment": to raise a newborn infant in isolation; but "by our very study of man, the knowledge of him is put out of our power"—nobody would be so cruel as to do such a thing. Such a child might, he thought, show how the true signals of sentiment emerge in a creature that had never received them.

The eighteenth century was a vintage era for "wild children," those raised—metaphorically or otherwise—by wolves. Linnaeus classified them as *Homo ferus*—wild men—whose nature would reveal what made *Homo sapiens* different. Most of the supposed examples were fakes, but a few were not.

In 1797 a young boy was found alone and almost naked in the forests of the Aveyron, in south-central France. He was captured, but he soon escaped. He was recaptured but the lad managed to escape a second time, but after a time he emerged from the woods under his own volition. He was about twelve years old, unable to speak, and savage in his behaviour. A vicious scar on his throat hinted that his parents had tried to kill their child. The boy appeared to have been without contact with others for almost his whole life and showed no obvious joy, fear, or gratitude when at last he met members of his own species. Here, perhaps, was an opportunity to investigate the springs of emotion.

A young student, Jean Marc Gaspard Itard, heard the story and saw the chance to test Rousseau's ideas. He took the forlorn youth to Paris and set out to raise him to the spiritual level of his fellow citizens.

Itard had trained as a tradesman but took up medicine at the time of the Revolution (he later became a pioneer in the study of diseases of the ear, nose, and throat). In contrast to Rousseau, he was convinced that the essence of the human condition lay in the ability to read the thoughts of others and, armed with that talent, to build a society in which passions could be kept in check for the good of all. His *Historical Account of the Discovery and Education of a Savage Man* sets out his theory that "MAN can find only in the bosom of society the eminent station that was destined for him in nature, . . . that moral superiority which has been said to be natural to man, is merely the result of civilisation."

The doctor took young Victor—whom he named after one of the few sounds, "o" (as in *eau*, the French word for water), he was able to recognize—into his household and set out to train him to express and respond to inner feelings. He was soon disappointed. The lad was "insensible to every species of moral affection, his discernment was never excited but by the stimulus of gluttony; his pleasure, an agreeable sensation of the organs of taste; his intelligence, a susceptibility of producing incoherent ideas, connected with his physical wants; in a word, his whole existence was a life purely animal."

Itard laboured for five years with both kindness and cruelty (the latter based on his charge's fear of heights) to transform the youth from monster into Frenchman, but with little success. Victor's behaviour stayed strange: he was obsessed with the sound of cracking walnuts but ignored gunshots close to his ears and loved to rock water back and forth in a cup. He never learned to speak and showed no gratitude for food or shelter. The sole sign he made of any response to the sentiments of others was a single attempt to try and comfort Itard's housekeeper after seeing her in tears following the death of her husband. Otherwise he stayed apart.

His protector insisted that the young man's failure to adapt to the inner worlds of those around him, and to express his own, arose because he had been rescued too late to pick up the skills needed, but that view was too optimistic. The lad would nowadays be diagnosed as deeply autistic, unable to respond to or give the signs—the smiles or frowns, or conversations—that bind people to the community in which they live. The dire effects of the illness show how our response to the sentiments of others makes us what we are.

The term *autism* was invented in the 1940s to describe a condition in which children fail to interact or to express sentiments apart from anger. They speak with difficulty or not at all and are filled with obsessions about particular foods, places, or clothes. About a third of them also suffer from epilepsy. Three out of four of those most severely affected struggle to cope with the condition throughout their lives. Victor had a grave form of the illness. Autism shades from the severe disturbance shown by the Wild Boy himself, through Asperger's syndrome (in which the language problems are less severe), to general problems in the development of normal conduct. Often the problem is noticed when parents become concerned by their child's depression or rage. The illness was once thought to be rare, with one child in two thousand affected. By the turn of the millennium the estimate was one in three hundred, but now the diagnosis is made far more often, with an incidence in both Britain and the United States of one in a hundred.

Autistics cannot understand the signals made by others and cannot make the full complement of their own. All children have that difficulty in their earliest years. As Darwin wrote in *Sketch of an Infant,* "No one can have attended to very young children without being struck at the unabashed manner in which they fixedly stare without blinking their eyes at a new face; an old person can look in this manner only at an animal or inanimate object. This, I believe, is the result of young children not thinking in the least about themselves, and therefore not being in the least shy, though they are sometimes afraid of strangers." In most infants such self-absorption passes, but an autistic child is locked into that phase for life. Many, when they look at other people, ignore the eyes, the flags of sentiment. They are equally unconcerned when someone else gazes long and hard at them.

The Expression of the Emotions used the blush as a prime example of a social cue, but embarrassment plays a lesser part in life today. Yawns—unacceptable in a nineteenth century drawing room—are nowadays more frequent. Many creatures open their mouths in such a gesture. Dogs do it, lions do it, even babies in the womb do it—but nobody really knows why. Theories abound. We open wide when we are tired, bored, or hungry. Perhaps a sudden drop in blood oxygen, or a surge of carbon dioxide pumped out by a tired body, is responsible—but breathing air rich in that gas, or with extra oxygen, makes no difference. It happens on hot days more than on cold, which leads to speculation that the action cools the brain. Whatever else a gaping mouth might do, yawn and the world yawns with you (and even reading about it can trigger the occurrence, as about half the readers of this book can now attest). It is a powerful and contagious signal, and *Emotions* discusses its use as a threat by baboons. In chimpanzees, too, it is sometimes a statement of sexual dominance (and, bizarrely, certain antidepressant drugs spark of bouts of yawning accompanied by sexual fantasies in women).

In normal children, the habit begins around the age of six. Such a spontaneous response to a second person's signal of mood is an unmistakable sign of empathy, of an ability to understand and to react to someone else's state of mind. Not, though, for children with autism, who are far less likely either to yawn or to emulate those made by people around them. The speed and extent with which a person yawns in response to another's involuntary gape may be a quick and objective measure of the degree of empathy to which he or she might be blessed. It is, as a result, sometimes used as a simple diagnostic tool for the condition.

Psychologists talk of "theory of mind," the ability to infer the mental state of others. People with autism have no insight into the inner world of their fellows and cannot express their own internal universe in a way that makes sense to those around them. They are blind to the messages written on another's countenance and find it hard to decipher gestures of anger, fear, sadness, or joy. Like chimpanzees (but unlike dogs), some autistic children cannot understand what is meant when their parent or doctor points at an object: they lack even that simple social talent.

Autistics also find it harder to tell people apart or to recognize a photograph of themselves. A certain group of brain cells is activated when monkeys or men see or copy the movements of others or observe an expression of pain, fear, or disgust. The same cells are also involved in the shared response to a yawn or a smile. These mirror neurons, as they are called, are almost silent in children with severe autism. Perhaps they are part of the system that helps us see into the souls of those around us. Their failure condemns those affected to a world whose other denizens act in mysterious and unpredictable ways.

There is no cure for the condition, although some symptoms, such as insomnia or depression, can be treated. The illness is four times more common in boys than in girls but shows no fit with race, social class, or parental education. Infection, immune problems, vaccines, heavy metals, drug use during pregnancy, Caesarian births, and defective family structure in Freudian mode (the child psychologist Bruno Bettelheim spoke of "refrigerator mothers") have all been blamed, but those claims do not stand up. Some say that the brain of a typical autistic child grows too fast too soon but then slows down. The amygdalae—those detectors of fear—are unduly active in some patients, but many other parts of the brain have also been implicated. Problems with serotonin, that universal alibi for disorders of emotion, may be to blame, and some autistic children synthesize the stuff less readily than normal. Certain drugs used against depression can help, as a further hint of a tie between social isolation and the emotional universe.

Genes are without doubt involved in some patients (although just one case in ten can be ascribed to a definite genetic cause). If one identical twin has autism, its sib has a seven-in-ten chance of contracting the disease, while the risk for nonidenticals is far lower. The incidence increases by a factor of twenty in the brothers and sisters of those with autism, and some among them who are not diagnosed with the disease nevertheless exhibit symptoms such as tactlessness, aloofness, or silence. A wide variety of genes has been implicated, but none contributes more than a small amount to the overall susceptibility to the disease. One common feature is a change in the numbers of copies—either through a new mutation or by inheritance—of particular segments of DNA scattered all over the genome.

Such behaviour sometimes presents itself as part of a larger medical problem. Fragile X syndrome is the commonest cause of mental disability among boys. It comes from a huge multiplication of a short section of DNA upon the X chromosome. Some patients have symptoms quite like those of autism (and some individuals diagnosed as autistic may in fact have fragile X). Other deletions, duplications, or reversals of a segment of chromosome are behind other cases of the illness. Other autistics have problems with a gene involved in the transmission of impulses between nerves. Yet more may have errors elsewhere in their DNA, and dozens of genes have been blamed. One candidate belongs to a group of genes that are multiplied in number in humans as compared with all other mammals and are active in the brain. Despite such hints, the biology of autism remains obscure, and the condition may be not one disease but many.

Autistic children are an experiment in emotion. Their isolation is mental rather than physical; they are cut off from the flow of information that passes among others. A world full of autistics could not function, for society depends on a silent dialogue in which every member's intentions are overtly or otherwise expressed. Civilisation turns on the ability to bear another's company.

Those who break its rules must be punished; and part of that involves the manipulation of the emotional state. By their nature prisons are places in which social interaction is reduced. Solitary confinement is a penalty far more severe than mere imprisonment, for it is autism imposed: a denial of what it means to be human, inflicted upon someone who once experienced a full range of feelings. The penalty is bitter indeed and is much employed by punitive societies, from mediaeval England to the modern United States. Charles Dickens visited a penitentiary in Philadelphia and wrote that "I hold this slow and daily tampering with the mysteries of the brain to be immeasurably worse than any torture of the body." The infamous "supermax" prison at Marion, Illinois, built to hold violent offenders together with political prisoners such as Black Panthers and members of the American Indian Movement, allowed almost nobody out of their cells for twenty years, even for exercise. It closed in 2007, but some of its more than forty replacements are just as brutal, feeding their inmates the tasteless prison concoction "Nutraloaf" to further reduce their contact with the world of the senses. Many inmates—like autistics—become anxious, agitated, and angry, and may end up insane, killing themselves should the chance arise.

If Zacarias Moussaoui, sentenced to life in solitary confinement for his supposed ties to the September 11 attacks, were allowed reading material in his soundproofed Colorado cell, he might learn something from both Dickens and Darwin about why he feels such hatred for those who do not share his views. As books are not available, he may wish instead to spend his solitary hours contemplating the expression of a condemned prisoner as the electricity passes through his head, which is said—in an echo of the great naturalist's own observations—to be one of uncontrollable horror.

CHAPTER 4

The Triumph of the Well Bred

CHARLES DARWIN WAS WORRIED ABOUT HIS PLANS FOR MARRIAGE. Perhaps the whole idea was a mistake because of the time diverted to family life at the expense of science. His diary records how he agonised over the pros and cons of matrimony, and his decision: "Marry, marry, marry!" In the end he did.

His spouse was his cousin, Emma Wedgwood. In falling for a relative he stuck to a clan tradition. The Darwins, like many upper-crust Victorians, had long preferred to share a bed with their kin. Charles's grandfather Josiah Wedgwood set up home with his third cousin Sarah Wedgwood. Their daughter Susannah chose Robert Darwin, Charles's father. Charles's uncle—Emma's father—had nine offspring, four of whom married cousins. The great evolutionist's own marriage was, in the end, happy, with ten children (and when his wife was in her early forties he wrote that "Emma has been very neglectful of late for we have not had a child for more than one whole year"). Even so, in Queen Victoria's fecund days the Darwin-Wedgwood dynasty did less well than most, for among the sixty-two uncles, cousins, and aunts (Emma and Charles included) who descended from Josiah, thirty-eight had no progeny who survived to adulthood.

Six years after his wife's last confinement, Darwin began to think about the dangers of inbreeding. His concern was picked up from another of his cousins, Francis Galton, the founder of eugenics, who had pointed out the dangers of marriage within the family.

Charles was anxious about his children: his tenth and last, Charles the younger, died while a baby; he was "backward in walking & talking, but intelligent and observant." Henrietta had a digestive illness not unlike her father's and took to her bed for years, and he feared that his son Leonard was "rather slow and backward" (which did not stop him from marrying his own cousin and serving as president of the Eugenics Society), while Horace had "attacks, many times a day, of shuddering & gasping & hysterical sobbing, semi-convulsive movements, with much distress of feeling." His second daughter, Elizabeth, "shivers & makes as many extraordinary grimaces as ever." George's problem was an irregular pulse, which hinted at "some deep flaw in his constitution" and, worst of all, his beloved Annie expired at the age of ten, throwing her parents into despair. As he wrote, "When we hear it said that a man carries in his constitution the seeds of an inherited disease there is much literal truth in the expression." At one point the naturalist even wrote to a friend that "we are a wretched family & ought to be exterminated." Might his illness and those of his sons and daughters be due to his own and his ancestors' choices of relatives as life partners? Was inbreeding a universal threat?

Darwin's first statement of concern came three years after *The Origin*, as an afterword to his book *On the Various Contrivances by Which British and Foreign Orchids Are Fertilised by Insects, and on the Good Effects of Intercrossing*. The last paragraph of that hefty work, most of it devoted to botanical minutiae, ends: "Nature thus tells us, in the most emphatic manner, that she abhors perpetual self-fertilisation. This conclusion seems to be of high importance, and perhaps justifies the lengthy details given in this volume. For may we not further infer as probable, in accordance with the belief of the vast majority of the breeders of our domestic productions, that marriage between near relatives is likewise in some way injurious,—that some unknown great good is derived from the union of individuals which have been kept distinct for many generations?"

The idea that children born to related parents might be in peril was already in the air. The first study of its risks came in 1851, when Sir William Wilde (father of Oscar) found, in work years ahead of its time, an increased incidence of deafness among the progeny of cousins. Sir Arthur Mitchell, the Deputy Commissioner in Lunacy for Scotland, had earlier claimed that the inbred fishing communities of northeast Scotland had an average hat size of six and seven-eighths, a quarter-inch less than that of their more open-minded agricultural neighbours—proof, he thought, of the malign effects of the marriage of kin upon the mental powers.

Sex within the family has a venerable history. The pharaohs lived through generations of the pastime in an attempt to preserve the bloodline of a god. Akhenaton, who lived around 1300 BC, first married his cousin Nefertiti, then a lesser wife, Kiya, then three of his own daughters by Nefertiti, and then (possibly) his own mother. The story is confused by difficulties with identifying quite who was who (and one of his supposed wives was in fact male), but incestuous affairs were without doubt common in ancient Egypt. Cleopatra herself may have been the scion of ten generations of brother-sister unions. The practice is condemned in Leviticus, where the children of Israel were enjoined that "after the doings of the land of Egypt, wherein ye dwelt, shall ye not do."

The belief that the children of cousins are bound to be unfit (and the desire of all governments to control their citizens' lives) still fuels a jaundiced view of the joys of sex within the household. In 2008 a British government minister, referring to the Pakistani population of Bradford, made the quite unjustified claim that "if you have a child with your cousin the likelihood is there will be a genetic problem." Many of his fellow citizens share that vague Galtonian sense that inbreeding is harmful, although their alarm rests on anecdote rather than on science.

All states are interested in how their subjects behave in the bedroom. For years England based its marital rules on those of the Church of England, which in turn descended from those of the Israelites. In 1907, after hundreds of hours of parliamentary discussion, the statutes were at last clarified. The new legislation removed some absurdities (such as the biologically meaningless law that forbade a widower to marry his dead wife's sister), but it also firmed up the prohibition against sex with close kin, be it father with daughter, or brother with sister.

Politicians often act on the basis of prejudice. Darwin did not. When faced with a scientific question—about sex or anything else—he set out not to speculate but to discover. To learn more about inbreeding he turned again to plants. *The Effects of Cross and Self-Fertilisation in the Vegetable Kingdom,* which appeared in 1876, describes a series of experiments on a wide variety of hermaphrodite plants forced to mate with themselves. Its verdict was clear: "The first and most important of the conclusions which may be drawn from the observations given in this volume, is that cross-fertilisation is generally beneficial, and self-fertilisation injurious." It was "as unmistakably plain that innumerable flowers are adapted for cross-fertilisation, as that the teeth and talons of a carnivorous animal are adapted for catching prey." Outbreeding—the exchange of genes between unrelated individuals—was the rule, and self-fertilisation an expensive exception. What was true of plants must, he thought, apply to animals, men and women included.

Flowering plants have sexual habits more imaginative than our own. *The Loves of the Plants*, a poem by Charles's grandfather Erasmus, is a work of science in two hundred pages of Arcadian verse. Many lines deal with the balance between male and female interests ("Each wanton beauty, tricked in all her grace, Shakes the bright dew-drops from her blushing face; In gay undress displays her rival charms, And calls her wandering lovers to her arms"—in other words, this species needs a pollinator). His grandson asked deeper questions in plainer prose. He found that many plants live in a reproductive universe that would have shocked the shepherds of Arcady. They have a system of choice that transcends the familiar preferences of one gender for its opposite. Most retain their original nature as hermaphrodites, with male and female parts in the same flower. They span the range from obligate self-fertilisers to others that make an absolute demand for pollen from another individual. Many among the latter group have imposed additional and refined laws of sexual choice upon their mates to ensure that they do not accept genes from their close kin.

The cross-fertilisation book was a first step in the scientific study of sex. Fifteen years earlier its author had noted that "we do not even in the least know the final cause of sexuality; why new beings should be produced by the union of the two sexual elements, instead of a process of parthenogenesis." Not much has changed. We are still not certain how the habit persists in the face of its obvious drawbacks in terms of cost, stress, and more—in Dr. Johnson's words, the expense damnable, the position ridiculous, and the pleasure fleeting. Parthenogenesis—virgin birth—guarantees that all the genes of those who indulge in it reach the next generation. It seems the obvious solution but remains rare—a few lizards and fish, and about one species of flowering plant in a thousand. Why mate when virgin birth guarantees your own genes a chance of survival? For a hermaphrodite, a bout of sex with oneself also ensures that the DNA is not diluted with that of an unrelated individual, and for creatures with separate males and females, incest is almost as effective at keeping genes in the family; but that too is widely frowned upon.

Plants hint at the answer. Some propagate themselves with shoots or broken fragments while plenty more have, like men and women, separate sexes. The majority of the flowering kinds have taken a lesser step towards asexuality, for they bear male and female functions on the same individual.

In spite of the chance they have for sex with themselves, many hermaphrodites insist on exchanging genes with a stranger. Others are happy to self-fertilise; they accept genes from a different flower on the same plant, or they evolve flowers with both male and female structures that can fertilise themselves. That pattern marks a real step towards the abandonment of sex.

Animals, too, have often tried to give up that expensive habit. Some, like a certain praying mantis and a Californian lizard, are true parthenogens, while a few, such as snails and worms, are hermaphrodites. Others go in for bizarre forms of close mating. The habits of mites would astonish any pornographer. A certain parasite of locusts gives birth to two types of male. The first clambers back into his mother and fertilises her. With his help she then produces a second brood, with a few males included—and those males then copulate with their sisters. In another mite, a mother mates with her grandson, the scion of her own daughter, the daughter herself a child of the mother and a son. Other mites confine themselves to brother-sister pairs—but they copulate before they are born.

Darwin saw that when it comes to sex, plants are better subjects for experiment than animals. Self-pollination in hermaphrodites marks a biological indulgence more extreme than the copulations of cousins that so concerned him, or sex between sons and mothers, fathers and daughters, or brothers and sisters. He set out to explore how often it took place and what it did to the health of his subjects, and, with luck, to learn something about the importance of that form of reproduction in general.

Within a few months of starting work as a planned pollinator in the greenhouse at Down House, he found that the effects of inbreeding could be dramatic. With the help of a botanical condom—a fine mesh to keep out insects—and a small paintbrush he could himself, like a bee, move male cells to the female parts of a flower and arrange that the plant received its own genes or those of another individual.

First, he noted that certain species would self-fertilise, while others refused to do so even when obliged to try. Among those that did, he soon discovered—somewhat to his alarm—that the habit did seem to damage later generations. His initial experiments were on toadflax, a common yellow-flowered weed. In the wild, outcrossing was the rule. In the greenhouse, Darwin could force his subjects to self-pollinate. He found a large and unexpected effect upon the next generation. The progeny of such crosses were smaller and less vigorous than those of plants allowed to mate with another. At first he supposed that his inbred offspring were weak because of some disease, or because they were grown in unsuitable soil. That was not so; however well they were treated, they stayed feeble. Darwin ran through a variety of species—carnations, tobacco, peas, monkey-flowers, morning glory, foxgloves, and others. With statistical help from Francis Galton he discovered that, almost without exception, those grown from crossed seed were taller, healthier, and more productive than were those from self-fertilised. Some experiments went on for several generations, and the effects of within-family sex often got worse with time. The inbreds suffered most of all when life was hard: when they were crowded, had to compete with their outcrossed kin, or were moved from the greenhouse to the rigours of the open air. The malign influence of selfing applied almost as much to species that went in for it in nature as to those that almost never did.

Once or twice, out of thousands tested, the descendants of a selfed individual were healthy and vigorous. Sometimes they even outgrew their competitors. One selfed morning glory (a species normally damaged by inbreeding) he referred to as "Hero" because its line flourished under that mode of reproduction. Why, he did not understand.

Even so, in most plants in the wild, to mate with a close relative appeared to be impossible, an expensive error, or at best a doctrine of last resort. The implications for humans were, perhaps, alarming. Darwin's view was, we now know, too inflexible. Almost all hermaphrodite plants have at least the potential to fertilise themselves, and many do so as a matter of course. (He also denied the importance of selfing in animals and was again mistaken: I myself once worked on hermaphrodite slugs, who manage perfectly well with sex within their own skins.) Only about one plant species in five prefers to self as a general habit. Other hermaphrodites that normally outcross will do so when no alternative is available, and only a minority avoid the pastime altogether. The notion that plants could be divided into two distinct types based on reproductive habits, which grew up in the century or so after his plant fertilisation book, is wrong. Hermaphrodites go from obligate selfers to determined outcrossers, but most are happy to adopt either practice as conditions change. Some shift from sexual to asexual in different places or as the seasons move on. For others, self-fertilisation is a side effect of sex, a kind of green onanism when a bee, as it flits from flower to flower, fertilises one flower with pollen from another on the same individual.

The decline in fitness of selfed individuals and the occasional appearance of healthy lineages both emerge from the simple rules of inheritance and from the existence of large amounts of hidden genetic damage in most plants and animals. A hermaphrodite that bears two different versions of a particular gene—a pea in which the DNA that codes for shape has the instructions for round seeds paired with another set for wrinkled—will, after self-fertilisation, produce half the next generation with the round-wrinkled mix, a quarter with round alone, and a quarter with just wrinkled. If they are again selfed, the pure round and pure wrinkled plants will have offspring identical to themselves, while those still bearing both kinds of instruction will repeat the proportions that emerged in the previous generation—a quarter pure round, a quarter pure wrinkled, and half with one copy of each variant. As selfing goes on, a smaller and smaller proportion of the population retains both versions of the shape gene. Generations of such crosses hence lead to the emergence of "pure lines," in which every individual is identical; all with round seeds or all with wrinkled. The same is true for every other variant in each line. In time, each will contain different combinations of genes for seed shape, colour, height, and so on.

To make a pure line is not easy, for if the original population contained hidden variants harmless in single dose but harmful in double (as many do), its descendants pay the price as such variants are exposed in double copy. The more the hidden damage, the less the chances of success. The effect can be spectacular. In loblolly pines, native to the southern United States, only one egg in fifty fertilised by pollen from the same plant survives. As a result the creation of an inbred line becomes almost impossible.

Such lines are at the centre of modern agriculture. Thousands have been produced for use as crops or flowers. Wheat, rice, barley, tomatoes, and more—all have their reproductive lives controlled by farmers, and nearly all are the rare survivors from vast numbers of inbred lines that paid the price for their damaged genes as the program went on. The survivors were those who drew luckily in the sexual lottery.

Darwin's conversations with breeders told him that the first few generations of kin mating caused the effects to get worse—because, we now know, more and more harmful genes emerge in double copy. Now and again, as in the famous morning glory "Hero," an inbred wins because it inherits, by chance, genes that increase, rather than decrease, its ability to survive. Its descendants thrive and may, in time, be used in millions on farms or in gardens.

Farmers, consciously or not, have built on that observation, and the same is true in the wild. In some species, well-adapted inbreds emerge to cope with the horrors of Nature, causing millions of identical individuals to fill the landscape. The advantages of selfing may depend on how predictable life is. The practice is more common when the struggle to survive involves starvation or bad weather, conditions which return more or less unchanged each year. Perhaps, by chance, a set of genes emerges that deals well with food shortage or with cold. It then pays to stick to that well-adapted combination rather than to mix it with other genes during a spasm of sex. Selfing is more frequent in cold and starved northern forests, where life is short, or when few mates are available. For hermaphrodites whose pollinators are grounded by bad weather, or if the season is so bad that it becomes hard to make a decent flower, a shift to self-fertilisation makes sense, for the choice is between an unhealthy brood or none at all.

If—as in the tropics—the main enemies are parasites or predators, which use sex to shift their own tactics by scrambling up their genes, selfing does not work. To stick to a single strategy is to play poker repeatedly with the same hand against an opponent who reshuffles on each deal; sooner or later he will draw a winning combination, and the selfer will lose everything.

The rules that apply to plants work for animals too. When—in an echo of the Down House experiments—wild mice are paired brother with sister in the laboratory and the offspring are released into nature, almost none survives. Inbred animals do not often die of obvious genetic disease, but their parenthood can much weaken them. Song sparrows on a small island off Canada's west coast have been ringed for years and their pedigrees worked out. Those born to close relatives are at more risk of death in bad weather than are others. The island of Soay, in the Saint Kilda group, is famous for its native sheep, which have been there since Viking times. The creatures are filled with worms, and those with the heaviest burden suffer most of all when faced with a vicious Scottish winter. Lambs born to close relatives have more parasites, and are at more risk of death in storms, than are those that are the offspring of unrelated parents. Inbred animals purged with medicine survive the tempest as well as do their outbred kin—proof that the worms are to blame. The finches on the Galapagos also pay the price for sex with kin, as do wild shrews, red deer, seals, toads, bats, and more.

Endangered species, given the shortage of mates, are at particular risk. Planned parenthood can help. The Florida panther was once on its last legs, with many animals plagued with kinky tails and undescended testicles because of the exposure of genetic damage in the tiny and inbred population that remained. In 1995, eight female Texas cougars (a related species) were introduced, and now the natives, with the help of their relatives' genes, have fought off their tail and testicle problems and returned to genetic health. The situation in zoos is just as bad. A mere hundred or so Arabian oryx were left in the wild by the 1960s, and most were soon killed by hunters. Two males and a female were rescued, and a dozen or so additional individuals were already in captivity. The entire world population of two thousand, domestic and nominally wild, descends from those few founders—and the most inbred individuals still have fewer young than average.

If mating within the family is bad for plants and animals, what might it do to people? Many are convinced of its dangers, and many societies try to limit the practice. In the United States, cousin marriage is a criminal offence in eight states and in a further twenty-two is at least illegal (although the rules are often ignored). An attempt to ban such alliances was defeated in Maryland as recently as 2000. In a nod to the eugenic agenda, Wisconsin restricts them to couples in which the wife or husband is infertile or the wife is older than fifty-five. The traditions of particular groups can lead to a reluctant acceptance of difference; uncle-niece wedlock is allowed in Rhode Island but only for Jews, and male Native Americans in Colorado are permitted to marry their step-daughters even if the practice is (despite the lack of shared genes) outlawed elsewhere. Sweden, in stark contrast, is happy for half-sibs—children with one parent in common—to enter into marital bliss. The nation has even considered allowing brothers and sisters to marry.

It is, needless to say, impossible to carry out planned crosses with men and women, but Charles Darwin came up with another way to test the dangers of inbreeding. First, he tried to get questions about cousin matrimony included in the 1871 census. He pointed out that "the marriages of cousins are objected to from their supposed injurious consequences; but this belief rests on no direct evidence. It is therefore manifestly desirable that the belief should either be proved false, or should be confirmed, so that in this latter case the marriages of cousins might be discouraged." His request was debated in the House of Commons but thrown out as "the grossest cruelty," for it would cause children to be "anatomised by science" (and from a parliamentary point of view the issue was almost traitorous, for the Queen herself had wed her cousin). A query about "lunatics, imbeciles and idiots" was allowed but was dropped from the next census, a decade later, as most people refused to answer it. The great naturalist, annoyed by his failure to persuade Parliament to ask a scientific question, complained about "ignorant members of our legislature." His son George was even more scathing about "the scornful laughter of the House, on the ground that the idle curiosity of philosophers was not to be satisfied."

George Darwin set out to build on his father's work. From the records of

Burke's Landed Gentry and the *Pall Mall Gazette*, together with a circular sent to lunatic asylums, he worked out the frequency of cousin marriage in various groups. Such unions were, it transpired, twice as common among noblemen as among the proletariat. His enquiry to the superintendents of asylums as to how many of those in their care were the scions of related parents was, as they pointed out, unlikely to pay off because of the mental state of their charges. Even so, George found no increase in the level of inbreeding among the patients compared with that of the general population (even if the Deputy Commissioner in Lunacy for Scotland assured him that most of his nation's idiots were the children of relatives).

Following his mixed success with lunatics, the young man went on to study the inmates of Oxford and Cambridge colleges. He asked the boat-race crews—"a picked body of athletic men"—how many had been born of cousins. After a correction for a falsified return from the stroke of Corpus Christi College Cambridge, he found that there was indeed a slight shortage of such inbred individuals among top oarsmen compared with the general population. The same was true among sporting boys in the principal schools for the upper and middle classes. In both cases the numbers were small and the evidence not altogether persuasive.

Then George had an idea that might produce large amounts of information on the sex lives of his fellow citizens. He used surnames—an inherited character—to estimate the extent of marriage among relatives. Two people with the same name, particularly a rare one such as "Darwin" or "Wedgwood," are more likely than average to descend from a common ancestor. Indeed, Sir William Wilde had already found that the parents of deaf children had a higher than average chance of sharing a surname.

George Darwin himself did little with the idea, but surnames have now been analyzed by the million. They give a new insight into mating patterns. In most places, such labels pass, like the Y chromosome, down the male line. The fit between the two is real but for several reasons imprecise. Common names (including "Jones," which means "son of John") originate many times in different places. To confuse the issue further, children may be adopted into new families, or agree to change their name for testamentary advantage. Some people take up a new tag because they do not like the one they were given. Illegitimacy, too, is a problem. These frailties weaken the link between shared names and shared genes.

Even so, a random set of a hundred pairs of British men showed a real tendency for those with the same surname to have common sets of genetic variants on the Y. Two who bore the same rare tag were far more likely to share a Y chromosome than were two who shared a frequent name. Names hence provide insight into genetic history (and the police have sometimes tracked down criminals by using shared DNA to search for a possible surname).

The failure with the census and his son's ambiguous results from idiots and Etonians suggested to Darwin that perhaps the effects of human inbreeding were less dire than he had feared. Honest as always, he admits in his book that "my son George has endeavoured to discover by a statistical investigation whether the marriages of first cousins are at all injurious, although this is a degree of relationship which would not be objected to in our domestic animals; and . . . he has come to the conclusion that on the whole points to its being very small" (he also removed his comment on the harmful effects of inbreeding from the second edition of *Orchids*). The effects of cousin marriage on health were too minor to be picked up by his observations, but the mass of information now available from official records, from surnames, and from patterns of shared genes shows that its influence cannot be ignored.

European aristocrats, like those of ancient Egypt, have long married their kin. In a world in which nobody mates with a relative, and with the births of parents and children separated by twenty to thirty years, each reader of these pages could have had, at the time of Charles Darwin's birth seven generations ago in 1809, a hundred and twenty-eight different ancestors—two multiplied by itself seven times. For almost everyone that figure is too high, since some marriage among kin is inevitable and usually involves people who have no idea that their spouse is a distant relative. Social pressure for matrimony within the household can further reduce the number of ancestors. Alfonso, the Infante of Spain who died in the 1960s, had just twelve ancestors seven generations back. King Alfonso XII, a contemporary of Darwin, had sixteen, while others in that noble line had between fifteen and twenty great-great-great-great-grandparents—far fewer than expected in a sexually open society. The Spaniards still marry their cousins at rates well above the European average and have the most inbred villages on the continent, with Asturias in the north the most inward-looking place of all.

The habit also thrives elsewhere. Many of India's thousands of communities insist on alliances within their own group. The Dravidian Hindus of the south have encouraged the matrimony of cousins, or uncles and nieces, for two thousand years, and continue to do so, with one marriage in ten between a man and his brother's daughter. The practice is also common in the world of Islam. In parts of Africa and the Middle East, a fifth of all alliances—in some places even more—are between close relatives. The tradition is not, contrary to popular belief, central to that religion but has become part of its culture. The Prophet discouraged the idea, but he did marry his daughter Fatima to her cousin Ali, which in the eyes of the faithful legitimises the habit. In Bombay one Hindu wedding in fifteen is between close relatives, but for Muslims the figure is three times as high. Among the nomadic Qashqai people of Iran the incidence of cousin marriage is three in four.

The picture in the West is quite different. Most of us rarely meet our relatives, let alone sleep with them. In most places fewer than one marriage in a hundred is between cousins. The English have long been among the least inbred nations in Europe. George Darwin had imagined that farm labourers "would hold together very closely" but was surprised to find how few relationships there were among kin in the rural districts of his time. England has never had a peasant class that sits on its own land for centuries and falls for the girl next door for lack of anything better. John Bull was a tenant. His master often forced him off, providing, as an unexpected consolation, a wider choice of mates. In modern Britain, husbands and wives are, on the average, sixth cousins. They are likely to share no more than one great-great-great-great-great-grandparent in common—someone who lived before George's father was born—and most have no idea that they are related. Only on islands (real or metaphorical) is cousin marriage common. Many Northern Irish travellers and a fifth of the people in parts of the Hebrides choose a first or second cousin as a mate. Faith is a better barrier than distance. In Bradford, the Pakistani community is among the most inbred in the world, with its British-born children even more likely than were their parents to marry a close relative (often still in Pakistan).

Social pressures can much reduce the extent of inbreeding. In Wisconsin and other places, the prospect of prison for sex with a cousin has, no doubt, ended many a romance. The geography of men and women (as seen in the patterns of Y chromosome and mitochondrial genes) shows that in much of Europe men have tended to stay at home while their mates come from elsewhere. In most peasant communities, the sons inherit the capital and do not wish to move or share with a neighbour's family. They prefer wives from far away who are, as an incidental, unlikely to share their DNA.

Close inbreeding can—as Darwin had found in the greenhouse—impose a real burden. All human populations contain damaged genes, manifest only when inherited in double copy. The children of relatives may pay the price for that legacy. As Bagehot wrote: "It has been said, not truly, but with an approximation to truth, that in 1802 every hereditary monarch was insane," and inbreeding was at least in part to blame. Every reader of these pages carries a single dose of at least one gene that would be fatal if present in double copy. Most inborn flaws are hidden by normal versions of the same gene; one British child in twenty-five hundred is born with two copies of the damaged gene that leads to cystic fibrosis, but one Briton in twenty-five has a single copy. If relatives mate and if their common ancestor bore a defective piece of DNA, the chance that each partner will inherit that fault by virtue of shared descent goes up. Their offspring are then at increased risk of receiving two damaged versions.

The malign effects of sex within a closed pool were noticed long ago. Omer Ibn Al-Khatab, the second caliph and a direct follower of Mohammed, advised members of a certain tribe to marry out because he thought they had become weak and unhealthy through their habit of sex with kin. The risk became manifest in 1908 when Sir Archibald Garrod identified an inborn illness called alkaptonuria, caused when an enzyme that breaks down certain food substances is damaged, as a condition in which two copies of the altered DNA were needed to show their effects. Symptoms include dark ear wax, smelly urine, and, later in life, heart, skeletal, and other problems. The disease is rare, with just one case in every twenty thousand births, but Garrod found that more than half of his patients were the children of first cousins. The same is true for other conditions. In France, with one marriage in five hundred of this kind, the incidence of cystic fibrosis is seven times overrepresented in the progeny of such matings.

In an unfortunate coincidence, certain places with a lot of sex among kin—North Africa, the Middle East, and the Indian subcontinent—also have a high incidence of inherited blood diseases that protect against malaria. Sickle cell is carried in single copy by almost half the members of some African populations, and related errors are almost as common in other places. Each is dangerous when inherited in double dose.

In Saudi Arabia, where in some villages eight out of ten people are wed to a cousin, such diseases are common. A fifth of all admissions to the nation's children's wards are due to hereditary disorders. Many families are not aware of the dangers, and the devotion to that pattern of marriage remains. Doctors now advise those at risk to screen pregnancies in the hope that genetic conditions might be detected before the infants are born, and the government insists that engaged couples be tested to see if both carry a shared but hidden problem. No overt pressure is brought to bear, but the incidence of marriages within the family has dropped by a fifth in a decade.

Even in places without such high levels of inborn disease, children born to cousins die younger than the average. Such unions among Mormons—not an unusually inbred group—from the mid-nineteenth to the mid-twentieth century led to a notable increase in ill health. The effects became worse as the infants grew older (perhaps because, unlike the accidents of infection or starvation that killed many of the pioneers' babies, death in old age has a strong genetic component). Heart disease is also more frequent among the offspring of such matings. A study of half a million pregnancies in the modern United States suggests that the death rate of children of cousins is about 5 percent higher than average. The products of uncle-niece marriages, a pattern frequent in India and elsewhere, do even worse, and although incest (sex between sibs, or between parents and children) is rare, the infants that issue pay a high price. A German brother and sister, adopted at birth and strangers until adult, had four children, two of whom were severely affected. A study of thirty or so Canadian babies born to such parents suggests that almost half inherit some abnormality.

A subtle but marked effect of inbreeding has emerged in Iceland. Among a hundred and fifty thousand couples born between 1800 and 1965, partners who were close relatives had more offspring than average. Even so, the proportion of their own children who reproduced (and hence the number of grandchildren born to the pair of relatives) was lower than average for first and second cousins, in part because many of the first-generation progeny died before their time. Charles Darwin and his cousin Emma may have been testimony to that effect: seven of their ten offspring either died young or lived on but stayed childless. Close mating may be more harmful to a family's prospects than was once supposed.

Continued inbreeding leads to a decrease in variability within a family line. The effect extends across the entire genome and means that overall level of DNA variation is lower in people who emerge from a limited pool of ancestors. A scan along the double helix can, as a result, give an insight into the extent of inbreeding. It shows that many illnesses—diabetes, heart disease, and more—are more frequent, and more severe, among those so revealed to have a family history of sex with kin. In the northern English city of Bradford, some members of the Pakistani community are uniform in long stretches of their genome. They pay the price in terms of health, and even their general liability to infection goes up. In Darwin's day, childhood death came mainly from contagious disease. His beloved Annie died of tuberculosis (although the diagnosis was the obscure "bilious fever with typhoid character"), but her plight may, as he feared, indeed have been in part due to her parents' marital history.

Sex is, needless to say, more complicated in the bedroom than in the greenhouse, and the simple fit of health with kinship does not always hold. In some places, relatives marry to keep wealth within the household, which means that there must be some assets to hold on to in the first place. In parts of India and Pakistan, cousin wedlock is more common among the affluent than among the very poor, who have no such financial incentive. The effects of cash outweigh those of genes. As a result, the children of such matings are less, rather than more, liable than the general population to suffer ill health. As with sheep on Soay, the environment also plays a part. In the poor and embittered Japan of the 1940s, cousin marriage had a severe effect on infant well-being, but twenty years later that had almost disappeared.

For both plants and people, sex usually involves another party. Almost always, he or she must choose or be chosen from a pool of potential mates. This calls for hard decisions. Some are obvious: whites tend to marry whites, and blacks, blacks; the rich marry the rich and the tall the tall—and, it is said, men tend to choose wives who look rather like their mothers. Plenty of religions make it hard to find a mate outside the creed. Language, place of birth, education, and more also affect sexual choice. As a result, for any man or woman the number of possible partners is far smaller than it might be. The observation is familiar enough, but for all creatures biology sharpens sexual choices in ways both obvious and less so.

The fundamental question about sex is: why bother? The habit much reduces the number of potential mates. It imposes the simple rule that only some individuals—those of a different gender from oneself—are available for copulation. Self-fertile plants may circumvent that fiat, but sexual creatures (ourselves included) often make it more stringent. Evolution sets up laws that ensure that no longer can any male mate with any female (or vice versa) and that an individual of one sex will be accepted by no more than a fraction of those of the other. Sex, in effect, becomes more sexual than before.

Darwin discusses such issues in his book *The Different Forms of Flowers on Plants of the Same Species*, published a year after the volume on self-fertilisation. It begins with a simple tale; the story of the children of the village of Downe, who made necklaces from cowslips. They could, they told him, use only some of the plants, those with a long "pin" that protruded from the flower, through which they could thread the plants together. Other flowers, instead of a pin, had no more than a short protrusion called a "thrum" and were of no use as juvenile jewellery. The cowslip's close relative the primrose was much the same.

Pins and thrums are, Darwin found, an additional mechanism of mate choice. Female pin flowers were more willing to accept pollen from male thrums than from males of their own kind. The same applied in the opposite direction. Part of the effect is mechanical, for a bee covered in pollen from a pin plant is more likely to brush it off against a thrum plant as it enters in search of a reward. More subtle mechanisms are also involved.

All males need a female, but the cowslip asks for more. The flower's form is inherited—which means that the plants decide whether to accept another's advances on genetic grounds. Like can mate only with unlike. The second sexual filter reduces the chance of an encounter between genetically similar plants and acts as a precaution against inbreeding. Nature has come up with a trick to reduce the proportion of individuals with whom genes can be shared. She has in effect invented more sexes.

Darwin's experiments have grown into a science that shows how, both in primroses and in people, partners are chosen in unexpected ways and that such choices much decrease the prospects of gene exchange between those who have recent ancestors in common.

Thirty or so plant groups have evolved systems much like that of the cowslip and primrose. The owner of Down House found some that came in not two but three forms, each of which would accept a partner only from outside its own class. Now we know many more examples of such physical barriers to gene exchange. Some species produce flowers that are mirror images of each other and can cross only left to right and right to left. In a bizarre arrangement in tropical gingers, some individuals are male in the morning and female in the afternoon, while others prefer the opposite pattern. As a result they can exchange genes only with those whose reproductive clock is different from their own. Once again, the imperative is to avoid carnal relations with those like themselves—with kin.

In his self-fertilisation experiments, Charles Darwin had no real idea of how and why his experimental subjects accepted or rejected particular kinds of pollen placed on their female organs. He referred only to the "extreme sensitiveness and delicate affinities of the reproductive system," which is poetic rather than persuasive. In fact, like the cowslip and the tropical ginger, they make extra tests of kinship before sex is allowed. The female parts compare the genes of the hopeful male cells with their own and reject pollen whose similarity is too close.

The process is hard at work in the many hermaphrodites that insist on outcrossing. Like sperm, a grain of pollen contains only a single set of genes, but unlike that potent liquid, a plant's male sex cell must fight its way through a barrier of female tissue to reach the egg. To do so it grows a long tube that penetrates into the appropriate part of its partner. Her protective layer bears the normal double complement of DNA in each cell. For her, choice is simple: compare the pollen with her own tissues, and if the two share too many genes, block it. For species that prefer to self-fertilise, the rule is relaxed or reversed.

For outcrossers, this system ensures that unrelated mates have the best chance of success. A new mutation in the identity cues carried by pollen is almost certain to succeed, for in its novelty it charms its way into the affections of all females, none of whom bear it themselves. As the generations go on, the new gene spreads—but it begins to lose its magic as more and more females inherit it and reject males with a matching copy. Each shift in male identity goes through the same process, and in time a system emerges in which almost every male has his own unique carnal calling card. That allows females to make decisions about the kinship of all hopeful males and to choose those least similar to themselves. They sometimes delay the decision, as the pollen is permitted to fertilise the egg, but embryos that come from mating with close kin are not allowed to develop.

In animals, too, the sexual examination can be rigorous. Just as plants choose what pollen tubes are allowed to grow, female insects may store the sperm of many males before deciding which should be allowed to travel further. A sperm within the female reproductive tract finds itself in a difficult and dangerous place filled with hurdles of many kinds. Promiscuous mammals have longer vaginas than do those who stick to a few mates. They make the male cells work harder to reach their goal. The vaginal tract is acidic, too, and sperm do not much like that. Of the millions implanted by a successful man, only a few hundred reach the neighbourhood of the egg, twenty or so make it to the point where they might be able to fertilise it, and just one gets in. After fertilisation, both plants and mice are happy to abort a high proportion of their embryos, particularly those that arise from the attentions of a closely related male.

Animals can force a potential partner to undergo a variety of additional identity tests even before copulation is allowed. Some are obvious, some much less so.

Simple familiarity can breed contempt. Unrelated Jewish infants brought up together in kibbutzim, or Asian children betrothed and made to live together when they are very young, may tend to avoid carnal contact when they grow up, and—in the latter case—are said to be less fertile and more liable than average to divorce. Brothers and sisters also tend not to fancy each other. Older sibs feel a stronger sense of aversion to their younger fellows than do the young to the old. The degree of kinship is the same, but the older children can be almost certain that the junior members of the household are the products of their own mother, for they saw them cared for as babies. A younger sib, on the other hand, knows only that an older individual lives under the same roof—which does not necessarily mean that they are close kin. They are less repelled by the thought of sex with somebody who might not, after all, be a relative. It takes fifteen years of shared residence for a younger brother or sister to build up the erotic revulsion that an older sib can generate by watching a few months of childcare.

Social pressures certainly play a part in our marital patterns, but genes are involved too. Some are obvious—people tend to marry someone of the same skin colour as themselves—but others, like those of the plants, are more subtle.

As a boy, I kept mice in my bedroom, a hobby quashed because of the awful stench. At the time that seemed no more than a nuisance, but in fact the aroma of mouse urine was an introduction to a new world of sexual communication. Quite unexpectedly, mice have more genes than we do. Almost all the extras are involved with the sense of smell. The genes that code for smell receptors—largely decayed in the human race—are in full order. Mice have hundreds, with overlapping sensitivities, which, together, can differentiate a vast diversity of scents. They help the animals choose their mates through the nasal passage.

Given the choice, an inbred laboratory female mouse prefers to copulate with a consort from a different line. So keen is she on a new swain that a pregnant female will resorb her fetuses to make herself available. Bedding soaked with male urine has the same effect. The females assess health as well as kinship. Their acute nostrils sniff out those who carry parasites and avoid them. Perhaps—as in the wormy sheep on the Isle of Soay—the healthiest males, with the most impressive (and expensive) statements of their fine condition, are less inbred or bear genes that help resist disease.

Mice live in an aggressive reproductive universe. Each male dominates a small patch in which he can monopolise the females, but his partners often hop over to a neighbour's territory. The male marks his boundaries with urine, and females base their choice on the same stuff; the more urine there is and the less familiar it smells, the better. Males are forced to engage in urinary battles in which each tries to water down his competitors' contributions. The females go for the most prolific and most aromatic. So potent is the identity cue that even the human nose, feeble as it is, can distinguish some inbred mouse lines by scent alone.

The perfume is based on a series of proteins, coded for by related genes in four different families, one of which has over a thousand members. Two others are expressed in a special organ with its own set of nerves, at the base of the nose. The proteins have a strong scent of their own and bind other male pheromones to make a cocktail of desire. The genes involved are so variable that—just as in flowers—females can choose a mate. They avoid potential swains with the same odour, and the same family history, as their own and also steer clear of those with low variability in the smell-related proteins, perhaps because their reduced and perhaps inbred state makes them less suitable as fathers.

Primates, too, signal with scent—which is why the aftershave industry does so well. Marmosets and tamarins, small New World monkeys, send out chemical messages containing dozens of constituents to mark their territories, to advertise when they are available, and to bond with their mates. A male's brain lights up in response to female chemicals when she is most fertile. The largest response is in those parts of the marmoset brain associated in humans with emotion.

We smell, as any marathon runner soon finds out. Bloodhounds can sniff out different individuals (and are confused by identical twins). Rats, too, can assess human kinship. The animals sniff for longer at an unfamiliar scent than at one they have already experienced. Give them a sweat-soaked shirt and they can tell whether they have smelt it before. When tested with the scent of the brother of a familiar subject, they sniff less than when given a sample from a cousin. To rats, at least, we have an aromatic identity.

But can men and women, like rats, mice, or marmosets, identify the sweet smell of the opposite sex? The case is not proved. Many of the human genes for odour reception have rusted away, leaving fewer than half the number at work in mice, and we lack the patch of sensory cells at the base of the nose that makes many other mammals more sensitive to scent. Generations of students have sniffed T-shirts worn by women at different stages of the ovulatory cycle, with inconsistent results. In spite of the undoubted genetic differences in individual ability to taste certain chemicals it has been hard to obtain clear results on the role of scent in human mate choice.

There are, nevertheless, hints that—like dogs around lampposts—men and women do pass romantic messages through the nose. Many perfumes contain synthetic musks of the kind used by monkeys or mice to choose a mate. One, a relative of testosterone, has long been touted as a chemical messenger. It is sold to farmers as "Boar Taint" to test the sexual receptivity of sows. Some people can smell it while others deny that it has any odour at all, but after several weeks of exposure even they begin to notice its presence, and the numbers of relevant receptors in the nose multiply greatly—in women more than in men.

As in plants, the smell of success lingers after the sex act is over. Human sperm pick up and move towards chemical signals from the egg, using a gene that sits right inside the group that codes for smell perception. The pore in the nose that picks up a scent molecule and the sperm cell membrane that responds to a signal from the egg both do much the same job and look remarkably alike—and, as befits their common heritage, some of the genes used by mice as they sniff the air to assess kinship are also active in sperm. In an unexpected link between two sexual worlds, the sperm receptor also responds to the scent of lily of the valley and, given the choice, will swim towards it. Whether human eggs prefer to attract, or to allow entry to, sperm genetically different from themselves, we do not yet know.

Mice, men, and flowers have converged in their mutual dislike of mating with a relative, but how did such cues of sexual individuality evolve? There are intriguing similarities between the mechanisms of choosing a mate and those that fight off infection.

All mammals, smelly or not, carry inherited identity cards on the surface of every cell. We do not accept kidney transplants because our immune system compares the donor's genes with our own, recognizes the tissue as foreign, and rejects it. The less related the source of the organ, the fewer genes in common and the lower the chance of success (which is why brothers and sisters accept each other's transplants better than do pairs of strangers). The whole system is based on a set of genes that sit close together on the DNA, in a section whose main job is to code for certain functions of the immune system, our prime defence against infection. Each comes in many different forms, and vast numbers of combinations are possible.

Disease is a potent agent of natural selection. Individuals with the most diverse set of immune-system genes, and those with variants not shared by many of their fellows, tend to fight off infection better than others. Mice and even fish seem to prefer to mate with those least similar to themselves in such genes, suggesting that the tie between such choices and disease resistance goes back a long way. In the battle against contagion, diversity pays, for the offspring of very distinct parents will have a new mix of defensive genes, confusing the parasites' ability to evolve fast enough to evade our immune system. It hence pays to choose someone as different from oneself as possible.

Darwin's work on the reproductive habits of plants has strayed into fields that would have shocked his contemporaries. His interest grew from his concerns about the effects of inbreeding in humans—on his own family in particular. Its influence is real (though less severe than he had predicted), and both plants and humans have evolved mechanisms that limit its effects. As had been the case in the race for nitrogen, when faced with the same challenges natural selection came up with similar solutions to very different parts of the kingdom of life. That would not have surprised him much (although he would, perhaps, be startled to discover that human sperm are attracted by the scent of lily of the valley).

The great man's concern about the possible damage done to his own children by inbreeding was not justified: of his sons, William became a banker and Leonard an army major. George was elected professor of astronomy and Francis a reader in botany at Cambridge, while Horace set up as a scientific-instrument maker and was for a time mayor of that city. The naturalist's offspring married into several eminent clans, including those of Keynes and Huxley, and in spite of their progenitor's concerns about inherited feebleness have produced dozens of descendants distinguished in science, medicine, and the professions. They stand as living proof that intellectual aristocrats, unlike their botanical and blue-blooded equivalents, need not pay the price of keeping their biological heritage in the family.

CHAPTER 5

The Domestic Ape

"LET THEM EAT CAKE!" SAID THE QUEEN, AND THEY DID. TWO CENTUries after the demise of Marie Antoinette, the poor are fat and the rich are thin. Across the globe, death from excess has, for the first time in history, overtaken that from deficiency. Eight hundred million people are hungry while a billion are overweight. The problem comes from evolution, as manipulated by man.

Darwin saw how farmers had bred from the best to produce new forms of life and used that notion to introduce the idea of natural selection. His argument is set out in the first chapter of *The Origin of Species*. Given time, with conscious or unconscious selection by breeders, new and improved versions of creatures—from potatoes to pigs and pigeons—will soon emerge. Were they to be found in nature rather than in fields, sties, or lofts, many would be recognised by naturalists as distinct species.

In *Variation of Animals and Plants Under Domestication*, published a decade after *The Origin*, Darwin went further in exploring the tame as the key to the wild. The book speaks of ancient times, when "a wild and unusually good variety of a native plant might attract the attention of some wise old savage; and he would transplant it, or sow its seed." That interesting event—the choice of favoured parents to form the next generation—was a microcosm of the process that had moulded life since it began. The variety of breeds seen on the farm, he wrote, was "an experiment on a gigantic scale," both a test of his theory and a proof of its power.

Savages have been replaced by scientists. Their work has produced many new varieties of plants and animals and reveals the eccentric history of the food on our plates. Modern biology has transformed farming. Planned breeding—directed evolution—has led to an enormous drop in the effort needed to feed ourselves. The British spend a sixth of their income on breakfast, lunch, and dinner and the Americans even less: a proportion down by half in the past five decades and by far more in the past five centuries. For most people, shortage has given way to glut. As a result, for many citizens of the developed world food is in effect free.

The blessings so brought are equivocal. The real price of sugar, starch, and fat has plummeted. Famine disguised as feast has spread across the globe. Evolution on the farm transformed society ten millennia ago and is doing the same today. The early farmers were powerful agents of selection on wheat, maize, cows, pigs, chickens, and more; but the influence of those domestic creatures on the biology of the farmers themselves was almost as great. It began as soon as the wild was domesticated, ten thousand years ago, and led people to evolve the ability to deal with new kinds of food. Today's shift in diet will, in time, have equally potent effects on the genes of our descendants.

A new global power is on the move. The empire of obesity began to flex its stomach in the 1980s and shows no sign of retreat. Twenty years before that dubious decade there was, in spite of a collapse in the real price of food, little sign of the coming wave of lard. Then, thanks to technology, came the industrialization of diet, the latest step in the scientific exploitation of the Darwinian machine. Now a tsunami of fat has struck the world, and the world is paying the price.

It does not take much to alter a nation's waistline. The rise in American obesity over the past thirty years can be blamed on an increase in calories equivalent to no more than an extra bottle of fizzy drink for each person each day. At the present rate, two-thirds of Americans and half of all Britons will be overweight by 2025, and Britain will be the fattest nation in Europe. Among industrial powers, only China and its neighbours are insulated from the scourge.

This twenty-first century plague is a side effect of the triumph of scientific farming—of planned evolution—and many of those afflicted suffer as a result of the inability of their own genes to deal with the new diet. Some will die young or fail to find a mate. As a result the obesity that emerges from modern agriculture will soon be—like the advent of farming itself—a potent agent of natural selection.

The people who laid out the first fields lived above the rivers that snaked across a green and leafy Levant. For millennia they had hunted game and gathered seeds, as humans had done throughout their history. Just after the peak of the last ice age, the Middle Eastern weather became wetter and warmer, and the grasses flourished. The gatherers prospered. Thirteen thousand years ago came a nasty shock, for the climate turned cold and dry for several centuries. The chill persuaded people to plant grains rather than just to collect them. Soon, the thermometer went up once more, the crops flourished, and agriculture made its presence felt. Within a few centuries, the Fertile Crescent was filled with tillers of the soil.

A similar way of life, based on maize and rice rather than on wheat and chickpeas, got under way in South America and China (and, in time, even in Papua New Guinea, where banana and sugar cane cultivation emerged sixty-five hundred years before the present). The habit spread fast. Farming reached Britain some five thousand years ago. The pursuit of wild game was more or less replaced by agriculture within just a couple of centuries, although people still ate plenty of seafood (and that remnant of the chase persists today). As new crops emerged, the locals began to husband animals that could feed on them. Soon, a hundred people could live on the space that before had supported but one.

The new economic system led to a grand simplification of diet. *Homo sapiens* has eaten some eighty thousand kinds of food since he first appeared on Earth. A dig in Syria of the homes of hunters who lived just before farming emerged turned up a hundred and fifty varieties of edible fruit, grain, and leaf in that single society. Even in the nineteenth century, Queensland aborigines feasted on two hundred and forty separate plants. With agriculture the cuisine became simpler. Within a few years, the Middle East had just eight crops: emmer and einkorn (antecedents of wheat), barley, peas, lentils, bitter vetch, and chickpeas. Quite soon, the people of the whole world considered together ate just half the number of plants once used by a single hunter-gatherer band. In most places just one or two crops—rice, maize, or wheat included—became the staple food. They kept that status for ten millennia.

Now things have changed once more. Some lucky citizens have taken a great leap backwards, to the hunter-gatherer diet. The middle classes now forage in pricey supermarkets for an eclectic range of edibles, from avocado to zucchini, imported from around the globe. The revolution of the rich began soon after Columbus, when exotic delicacies such as potatoes, peanuts, and tomatoes were brought from the New World. Other delicacies went the other way, albeit sometimes after a long delay: broccoli, for example, was scarcely known in the United States until the 1920s. On both sides of the Atlantic, those who can afford it have put ten thousand years of dietary history into reverse.

Advocates of avocado are still in a minority. Many of their fellow Americans and Europeans eat meals almost as dull as those of the first peasants, without the privilege of growing the raw materials themselves. Just as at the dawn of agriculture, their choices are narrower than were those of their parents and grandparents. Cheeseburgers, french fries, and sweet drinks are full of cheap energy, and the poor have seized upon them. Nowadays, the British and the Americans obtain twice as many calories from fats as did their immediate forebears. On average, the intake of sodium has gone up by ten times and that of calcium has fallen by half compared with earlier times.

The junk food revolution tells the tale of artificial Darwinism in all its details. The taming of the hamburger shows how man, the most domestic creature of all, has paid a high price for testing the biological limits set by his own evolution.

The first farmers, like the modern poor, became less healthy as their dietary options shrank. The symptoms were different from those of today, but the causes—an abundant but inadequate cuisine—were the same. Their bones show signs of deficiency disease, and the average height of adults dropped by six inches as the new way of life spread—a loss not regained for several thousand years. In North America, where maize, worshipped as a god, became the basis of almost every meal, another problem was a shortage of iron, for maize lacks that mineral and also interferes with the body's ability to absorb it from meat. Many people became anemic. No doubt they were tired, weak, and depressed as they pursued their wretched lives as tillers of the soil.

Deficiency and its diseases—lassitude, infirmity, and sadness included—have returned, but disguised as excess. Thirty thousand premature deaths a year in the United Kingdom are due to expanded waistlines, and that number is ten times higher in the United States, where, in 2005, obesity overtook smoking as the main preventable cause of mortality. As America's spending on food as a proportion of national income went down by half, its outlay for health care multiplied by three. In central and eastern Europe, even more healthy years of life are lost per head than in Britain. The present generation—those who grew up before the new age of edible trash—may be the longest-lived in history.

The problem for their children is fat. Medicine has long known how dangerous the blight can be; in Hippocrates' words, "Corpulence is not only a disease itself, but the harbinger of others." Thousands are dying before their time of heart disease, stroke, cancer, and diabetes, the four horsemen of the obese. Many others suffer from gout, arthritis, bladder problems, reduced fertility, and other conditions that affect the fat far more than they do the thin. To grow to resemble an apple rather than a pear is dangerous indeed, for three extra inches on the waist are much more harmful than the same on the backside—and around the globe the apples are replacing the pears, even among women, who used to put more inches on the bottom than on the belly. The apple brigade stores fat around the liver, which readily releases fat itself, hormones, and agents of inflammation into the bloodstream.

In the modern United States, as in the New World at the dawn of agriculture, Native Americans have paid a high price for the change in diet. A century ago, many kept to their traditional cuisine. The Pima Indians of Arizona—the Corn People, as they called themselves—stayed thin as they ate their hearty meals of tortillas or porridge, based on maize. Now they turn to burgers, chicken, and sweetened drinks. In fact, today's obese Pima eat just as much corn as their grandparents did, except that now the seeds have been through a cow, a chicken, or a soft-drink factory first.

Cheap corn gave birth to fast food. One American meal in five is eaten in the car; and the maize needed to provide each of its four passengers with a cheeseburger would more than fill its gas tank. A Chicken McNugget has thirty-eight ingredients, thirteen of which come from that crop. The fizzy beverage that washes it down is based on corn syrup, and the raw material of a milkshake came from a cow fed in a yard, on corn, rather than in a field, on grass. The "natural strawberry flavor" that adds its dubious tang to the shake is natural only in the sense that it is synthesised from corn and not chemicals. A quarter of the food items in American supermarkets now contain maize, and in these stores the rows of cheap packaged products—thousands are introduced each year—bear witness to an agricultural revolution that has taken place in the lifetime of most readers of these pages.

Seed crops—corn most of all—are good at turning sunshine into food. Even better, they are easy to store and move. Cows evolved to eat grass, but it now makes more economic sense to feed them grain on giant lots. More than half the maize and soy grown in the world is eaten by animals. As a result, global meat production has quadrupled since the 1960s, and the amount of flesh available per head has doubled.

Scientific farming has done in a few decades what took peasants centuries to achieve with no science at all, but their approach was almost identical to that of modern technologists. They understood little of what they were doing and initially may not even have made the tie between sex and reproduction. By the Middle Ages the idea that attributes ran in families was accepted; as the 1566 book *The Fower Cheifyst Offices Belongyng to Horsemanshippe* put it: "It is naturally geven to every beast for the moste parte to engender hys lyke." By then, artificial selection was well under way (even if the horse-racing and dog-fancying fraternities clung to the odd idea that qualities were passed down only the male line). In the eighteenth century, English improvers became aware of the need to mate animals of equal "beauty," and agricultural science began. Robert Bakewell, chief among the breeders, was frank about his motives. He called his barrel-chested New Leicester sheep "machines for turning herbage . . . into money" and hired out his rams for stud at a thousand pounds a season—a huge sum for those days.

Now animal breeding is a massive business. Champion bulls and stallions can sire thousands of offspring, and new statistical methods allow their young to be compared over hundreds of farms to see which have done best. Sometimes the actual genes involved are still not known: milk yield in cows has doubled since the 1940s, but the sections of DNA that did the job stayed hidden for sixty years. Molecular biology is changing that, with the DNA sequence of most domestic animals now complete, together with maps of hidden diversity that can track down where productive variants might be. Artificial aids—mechanical cows into which bulls can ejaculate and have their semen smeared across the globe, cloned sheep, engineered crops, and more—promise great things. The annual gain in meat or milk production brought by genetics is, in the developed world, around one and a half percent a year, over a billion dollars' worth in Europe alone. With an expected doubling of meat consumption in the next decade, even that may not keep up with demand. Plant technology has also been successful, as many genes for high yield or disease resistance have been tracked down or brought in from our domestics' wild relatives. Agriculture now works with foresight, a talent quite unknown to evolution but used, at least subconsciously, by the first farmers.

What did it take to become domestic? The basic demand is for a creature able to coexist with humans. Men can then choose, often without much thought, the most favoured individuals to found the next generation. Improvement becomes inevitable.

Darwin knew little about the origins of fruits, grains, and vegetables: "Botanists have generally neglected cultivated varieties, as beneath their notice. In several cases the wild prototype is unknown or doubtfully known. . . . Not a few botanists believe that several of our anciently cultivated plants have become so profoundly modified that it is not possible now to recognise their aboriginal parent-forms." Now the parents have been found, hidden in their descendants' DNA.

All crops have a lot in common. When compared with their wild predecessors, from tomatoes to barley and from chickpeas to plums, the tamed versions are less diverse, grow taller and less branched, and have fewer but larger fruits or grains that taste less bitter. They flower at different times of year and their seeds spring into life at once when planted rather than (like those of many of their wild fellows) demanding a long rest.

The tale of such changes is hidden in the double helix. The story of corn—the raw material of junk food—shows best how biology can reveal the past. Maize descends from a wild plant moulded into a dietary staple so different in appearance from its ancestor that for many years its origin was unknown.

The story of maize is that of the New World. Darwin himself knew that it was ancient, for on the *Beagle* voyage he found cobs embedded in a beach raised by slow upheaval many feet above the sea. Its story began in southern Mexico around eight thousand years ago, when people began to harvest, and then to grow, a wild grass called *teosinte* (the "grain of the gods"). Teosinte is still abundant over much of South America, although several of its dozen or so species are under threat. Male and female organs are held in different places on the same individual, with a "tassel" that bears pollen, and a number of small spikes that carry the female parts.

At first sight, teosinte looks quite unlike the familiar corn on the cob, and it was once assumed to be a relative of rice. A teosinte cob—a family of seeds held together on the same structure—is little more than an inch long. When mature, the seeds, each within its own hard coat, form an "ear," with half a dozen or so separate segments. The coat protects them from the digestive juices of the animals and birds that eat the cob. Each seed breaks off when ripe and, with luck, passes through the gut, falls on fertile ground, and germinates to form the next generation.

The corn cob of today, in contrast, has five hundred or more kernels. Its seeds are far larger, come in a variety of colours, and contain much more starch. They do not fall off without help and lack a protective outer sheath. They are, as a result, digested when eaten, rather then excreted. If at the end of the season the whole cob is not harvested but falls to the ground, it bears so many seeds that almost none survive the intense competition for light and food. Maize is hence entirely dependent on its human masters for reproduction.

Even so, the kinship of maize and teosinte is close enough to allow certain wild strains to hybridise with their tamed descendants (a talent that farmers hate, for it can degrade their crop, although scientists use it to rescue valuable genes before the natives disappear). The DNA of modern varieties is closest to that of the teosinte that grows in the hills around the Balsas river basin in southwest Mexico. There, McDonald's finds its roots. The oldest known cobs, sixty-three hundred years old, come from a cave in the valley of Oaxaca, four hundred kilometres away. At about that time, the people of Central America began to thrive on their tamed grass. They even learned to treat it with lime to release its essential vitamins (a talent forgotten until the mid-twentieth century, when the deficiency disease pellagra was tracked down to a diet of untreated maize).

At least a thousand genes in modern maize differ from those of teosinte. Fossil DNA from seeds forty-five hundred years old shows that the farmers had already selected genes to improve grain quality and size. Just five genes, or groups of genes, were responsible for much of the shift towards the domestic form, although many more play a part. The move from grass to food involved mutations that change the slim side branches of the grass into stout maize ears, others that remove the hard case around each seed, and others that ensure that the grains stick to the cob and do not shatter when touched. Long stretches of DNA on either side of those points scarcely vary at all, a hint that large blocks of inherited material were dragged through the population by breeders as soon as a new attribute was noticed.

Corn improvement is now an industry. The plant is the most widely cultivated in the world, with three hundred million tons grown each year in the United States alone. It has been mutated, selected, and hybridised to give hundreds of distinct strains. Some are seven metres tall and some much shorter; some are large, coarse, and used as cattle fodder, while others have tiny ears, the size of those of teosinte, just right for a cocktail snack. Sweet corn is full of sugar. The starch itself, in some kinds, bursts apart when heated to produce popcorn. The plant now flourishes from the far north to the tropics and is far more productive than its ancestors of even twenty years ago. The science of maize has changed the global economy more than has nuclear power.

The maize genome has a bizarre and unexpected structure. It contains almost as much DNA as our own cells do and boasts twice as many genes. Most consists of bits of mobile DNA that invaded long ago. Some of those molecular parasites can no longer copy themselves and sit sullenly in place, while others wake up now and again and move to a new site. They cause mutations as they go and may capture working genes, altering their effects. Many of the mutations involved in the improvement of maize emerged from this constant flux. Maize DNA still changes fast. Although some inbred lines descend from a shared ancestor that lived just a few decades ago, they are already as distinct from each other as are humans and chimpanzees. The mobile elements have been so active that when two inbred lines are compared, on average a fifth of all genes differ in where they sit on their chromosome. Corn is a plain food with a complicated biology.

Other crops have a less checkered history. Apples are easy. Fifty years ago, Almaty, in Kazakhstan, was—like Norwich in Tudor times—"either a city in an orchard or an orchard in a city." Its name means "father of apples," but the place is now more notable for its Porsche and Mercedes dealerships. The city and its surrounds were the site of a great domestication. The genes of the chloroplast—the green structure found in leaves—show that the apples we eat today are almost all the progeny of just two ancient Kazakh trees. Those mothers of all the world's apples grew not far from Almaty. Wild trees, some as big as an oak, are still scattered through the nearby Tien Shan Mountains. They are part of what was once a vast fruit forest filled with walnuts, grapes, and apricots as well as apples. Today's varieties, from the insipid Golden Delicious to rare strains such as Zuccalmaglio, have emerged through mutations and selective breeding in lines descended from those two progenitors. They are maintained with grafts and cuttings.

Unlike the small and bitter crabs borne by most wild apples, the fruits of the Tien Shan are large and sweet. They became luscious when the trees changed their reproductive partners. The seeds of most wild apples are moved by birds, which peck at the fruit and scatter seeds in their excrement, but in the Tien Shan, the Mountains of Heaven, bears do the job instead. A bird is happy with a small reward, but a hefty mammal demands a more substantial bait. The trees grew sweet apples to oblige. Eight thousand years ago, people and their horses moved into the fruit forest and developed a taste for the ursine delicacy. Kazakh apple seeds travelled in horse and human guts down the Silk Road that skirts the mountains. Now their descendants fill supermarkets across the globe.

The potato, like the apple, traces its origin to a small patch of land, in Peru, north of Lake Titicaca. It has been cultivated for five thousand years and has diverged into a large variety of forms. Lentils, peas, and chickpeas, too, also each descend from a single wild ancestor. The various strains of rice, in contrast, emerged separately from two or three distinct species of wild grass cultivated in China. Wheat is different again, for the modern crop emerged as a result of hybridisation between several species of grass (some still around today) that came together to generate a plant with many more chromosomes than before.

Not long after wheat, chickpeas, and the rest appeared on the plate, wild beasts were invited into the family. Only a few accepted the offer, and most did so before the time of Christ. Quite soon, society was transformed. Cows, pigs, horses, and sheep became every farmer's treasured possessions, and much of man's effort was devoted to keeping them happy. As soon as they abandoned the wild, the animals began to change, all in more or less the same way.

Farm animals, of whatever kind, tend—like their botanical equivalents—to follow some general rules. They are smaller and more lightly built than their unbroken counterparts, with shorter faces and smaller jaws. Often, they vary more in colour and shape, and many among them develop spotted coats. They grow fatter, with longer intestines, breed throughout the year, and make more milk. Males and females have less physical difference than they did in the wild. In wild cattle, horses, and pigs much of the force of natural selection involves sexual success. Battles among males led to the evolution of expensive horns or tusks, with days spent locked in combat. Once sex is under human control, that wasteful effort can be redirected to making milk, meat, or wool, which is why domestic bulls or rams are less infuriated by their rivals than are their untamed relatives. In their lazy lives they tend towards promiscuity rather than the faithful bonds some of their ancestors preferred—which is useful for farmers when they wish to choose superior animals as parents.

One species in particular was quick to abandon its ancestral habits. It was the first to accept servitude and has used its own personality to manipulate mankind. It reveals, more than any other, what it takes to become tame.

Darwin was a dog lover. He devotes the first chapter of *The Variation of Animals and Plants Under Domestication* to the animal's history. So great was canine diversity even in his day (and it has increased since then) that he was uncertain whether dogs had descended from a solitary ancestor, the wolf, or from several, with the fox and jackal as additional candidates (he did dismiss the widespread view that each breed had descended from a separate wild ancestor, now extinct). As he points out in *Variation Under Domestication*, even barbarians attend to the qualities of their pets, to such a degree that the dogs of Tierra del Fuego have gained the ability to knock limpets off rocks. The breeders were often ruthless: his book tells of Lord Rivers, who, when asked why he always had first-rate greyhounds, answered, "I breed many, and hang many."

As he noted, the dog is now the most varied of all mammals, in both mind and body. Some breeds were ancient. On a visit to the British Museum Darwin identified images of a mastiff on Assyrian monuments from the sixth century BC. Others were more recent and had diverged much more from their wild ancestor. Some attributes—such as the shape of the head and the receding jaw of the bulldog and pug—might, he thought, have arisen as sudden "monstrosities" (mutations, we would call them), but the majority came from the slow accumulation of favoured forms. The dog was a marvellous model of how flesh can be moulded by human choice.

DNA shows that all dogs are the descendants of wolves, which still cross with them when they get the chance (*Domestication* tells of "the manner in which Fochabers, in Scotland, was stocked with a multitude of curs of a most wolfish aspect, from a single hybrid-wolf brought into that district"). The earliest bones found with those of humans, in a German dig, are some fifteen thousand years old, and the animals probably loitered around campfires long before that—which means they were welcomed into the household well before any other creature. Even in their first days they changed, with shorter legs than their wolfish ancestors—a hint that they no longer roamed the countryside.

Since then, the animals have been subdivided into a vast variety of forms. Four hundred breeds have been named, and a hundred and fifty have official pedigree societies. They keep a close eye on their charges' sex lives, for the rules insist that both parents belong to a rigidly defined type and that any dubious bloodline be thrown out. Such exclusivity can lead to rapid change.

Some breeds (mastiffs, chows, and salukis included) have been distinct for centuries (although the pharaoh hound, with its pointed ears and short coat that resemble those of the images on Egyptian tombs, is fake: a modern copy of an extinct animal). Most breeds, though, are less than four hundred years old, and many even younger. Their vast diversity is witness to what human choice can do to a once wild animal.

In 1815 there were only about fifteen designated breeds in Britain. The first formal dog show was held in 1859, the year of *The Origin*, and by then the number of breeds had risen to around fifty. Many of the most popular of today's hundreds of varieties—terriers, spaniels, retrievers, and so on—trace their origin as distinct varieties no further than the past century, which means that they have gained an identity in just fifty or so canine generations. The genes show that nearly all were founded by fewer males than females, evidence that—in the ancient belief that quality passes only through fathers—a popular sire was mated with many bitches. Some males still father over a hundred litters, a pattern shockingly at variance with the monogamous sex life of the wolf. As breeders hold to their eccentric belief in the power of sperm over egg by choosing only the very best as sires, they much reduce the size of the male population available for reproduction.

Sometimes just one mutation can spark a new variety. The largest dog, the Irish wolfhound, stands three feet high at the shoulder and tips the scales at well over a hundred pounds. Sixty Chihuahuas would fit into one of those—but remarkably, the difference in size is due to a single gene, which comes in one form in the big animal and another in the small. The whippet is a racing dog. It too owes some of its identity to one genetic change. Now and again a heavily muscled individual—a "bully whippet"—appears in a litter. It bears two copies of an altered gene for a muscle protein and is grossly misshapen. In most cases such pups are killed at birth. Many other whippets carry just a single copy of that variant. Because they are faster than average, the gene was inadvertently selected for as the animals were bred for speed (and it has now revealed itself in beef cattle and even in a young German woman whose mother was a champion sprinter). Perhaps the most repellent of dogs is the Mexican hairless, or Xoloitzcuintli, originally bred for food (and used as a bed warmer by the Aztecs). As its name suggests, it is entirely bald. Again, a single mutation, in a gene that in humans leads to loss of hair and sweat glands, is responsible (and, unusually, ancient statues show that the error has been around for three thousand years). Darwin himself identified a family with the human form of the condition—and almost worked out the pattern of inheritance, for he noted that it was passed through daughters but expressed only in sons, exactly what is expected from its position on the X chromosome.

A few other dogs owe their distinctiveness to such simple inherited errors (the dachshund, for example, bears a mutation that in humans leads to dwarfism). For most named forms, though, divergence involves many genes. As the differences build up, each gains a unique identity. Canine diversity is arranged in a way quite distinct from our own. People, wherever they come from, are more notable for similarities than for differences, but a large part of the variability among dogs as a whole emerges from divergence among breeds. The pedigree clubs, by encouraging intense inbreeding within particular lines, have been real barriers to the movement of DNA. The three hundred thousand golden retrievers in Britain trace their descent in the past thirty years to around just seven thousand males. Other breeds have lost nine-tenths of their total variation as a result of close mating in just the dozen or so generations since the 1970s. The dogs have paid a high price. Determined—or deranged—insistence on forcing each line towards an arbitrary standard has led to King Charles spaniels whose brains are too big for their skulls and pugs whose eyes pop out so far that they get scratched whenever the animal bumps into something. Pugs are so inbred that the ten thousand in Britain share recent ancestry with only around fifty animals.

In spite of the damage they suffer, dogs have exploited humans very effectively for they do not pay with their lives, or the products of their bodies, for food and shelter. No other creature is so tied to its master, and no other domestic has been so subdivided. Most farm animals joined the family long after dogs did, and some walked into the fields several times in different places.

A hamburger has a complicated history. As the *Domestication* book notes, the cow was tamed on two continents, in Africa and in the Middle East. Cattle were precious long before they were farmed. The caves around Lascaux contain images of more than a hundred aurochs, the cow's gigantic ancestor. That impressive beast roamed wild in Europe, North Africa, and parts of Asia before dying out in the 1620s. Sumerians had a cow goddess, the "Midwife of the Land and Mother of the Gods." Later came the Semitic deity Ishtar, whose bull-god partner made enough semen to fill the Tigris. Egypt, too, had a bovine obsession. The pharaoh was called The Mighty Bull, with a tail on his kilt and bull legs on his throne. The Israelites for a time worshipped a golden calf (and suffered divine displeasure for their ways). The cults of the Minotaur, the toreador, and the Western show how the animal retains its emotional power—and some golfers, it is said, still use a desiccated bull penis as a lucky putter.

The bones of domesticated cattle appear in the Middle East around nine thousand years ago and in Europe around 5500 BC. Cows continued to mate with wild bulls for thousands of years. Ancient DNA shows that the female, mitochondrial, lineages of today's European cattle are quite distinct from those of their aurochs ancestors, while their Y chromosomes, the indicators of male ancestry, resemble the aurochs' male chromosomes more closely. Wild bulls must have continued to impose their desires upon domestic cows, with or without man's consent.

Bacon sandwiches tell a different tale. Pigs came into the household on several occasions, in different parts of Europe and the Near East, with some later input from Asia. Fossil DNA reveals a wave of Near Eastern pigs that moved into Europe and was then replaced by a second domestication of European wild boar. Horses, too, have several origins, with one centre close to the home of the apples in Kazakhstan. Traces of mare's milk (still popular in that nation) on fifty-five-hundred-year-old pottery fragments suggest that they were tamed by then. Sex bias by horse breeders has been extreme. Ninety-five percent of the three hundred thousand racehorses alive today bear the same Y chromosome as evidence of descent from a single stallion. He was the Darley Arabian, who was brought to England from Syria in 1704 by the British Consul, Thomas Darley. Two others, the Byerley Turk and the Godolphin Barb, provided nearly all the remaining male lineages. Europe has no native sheep or goats, and the domestic forms had a simple and single origin in the Middle East.

The chicken, whose rendered flesh is a staple of the junk-food diet, descends from two or more species of Asian jungle fowl—and is now, with a population of thirty billion, the commonest bird in the world. It spread from its native Thailand to fill Europe and the far Pacific, and from there to reach South America before Columbus. In spite of their long years on the farm, the birds retain a lot of diversity, perhaps because, like cattle, they continued to cross with their wild relatives. Their reproductive lives have altered more than those of any other creature, for some lay ten times as many eggs a year as do jungle fowl, each twice as heavy as those of their ancestors.

As they left the wild, the minds of those animals changed. The inmates of zoos submit but are not tame, let alone friendly, as many keepers know to their cost. Only those willing to bow to our desires have a chance of domestication. As Darwin put it, "Complete subjugation generally depends on an animal being social in its habits, and on receiving man as the chief of the herd or family." A hundred and fifty million Indian water buffalo live in harmony with their owners, but that creature's close cousin the African water buffalo has a strict hierarchy within its herds. Its pecking order does not give entry to men, and it has never been tamed. The African water buffalo remains one of the most dangerous of all mammals. Elephants are semidomestic at best and kill many people each year, and plenty of breeds of cattle are happy to turn on their masters. Sheep, too, show elements of the fragile personality of their ancestors, for they panic if disturbed. An aggressive animal is no help to man or beast, and farms need good behaviour. Men bred from the most submissive, leading farm animals to grow less alert, less active, and less angry than their ancient fellows.

Their good nature is coded for within the skull. The brains of domestic animals are, without exception, smaller than those of their ancestors—the pig's by a third and the horse's by a sixth. The hardest job for the first farmers was to persuade the wild to become tame, but once that Rubicon was crossed, the domestics could put aside their need to outwit Nature and pass the responsibility to us. Today they spend their time in a sort of calm and extended youth in which the trials of life are dealt with by their masters.

Nowhere is the importance of behaviour better seen than by the fireside. The wolf is aloof and suspicious. It avoids humans as much as possible. Dogs live in a very different mental universe. Given a choice of dog or man, a wolf cub will run to the dog, a puppy to the human. Wolves hunt in packs, while feral dogs live in chaotic and quarrelsome groups that soon split up. The canine mind has been modified in many ways. Men and women follow another person's gaze. Point at an object, and all eyes will turn towards it. Dogs share that talent, and if its owner indicates where a bone is hidden with a glance or a gesture, the animal runs to the right place. Wolves are baffled. So attuned are dogs to their masters' moods that they will yawn when their owner does.

Men become fond of their canine companions—who return the sentiment. When left with a stranger, a dog plays less than when with its master. The bond between man and pet can last for years, and Darwin noted how his own favourite responded to him at once after his long absence on the *Beagle*. Owners often delude themselves that their pet understands everything they say. That is not true, but the animals have without doubt some insight into the human mind.

To gain that talent, its brain has been much modified. Regions associated with aggression have shrunk, and there are changes to the hypothalamus, the bridge between the nervous and endocrine systems. As a result, our favourite pets (unlike wolves) breed year-round. The hypothalamus is a centre of activity for the neurotransmitter serotonin, which, when in short supply in humans, is associated with aggression, impulsive behaviour, depression, and anxiety. A deficiency of the chemical is also behind the attacks of rage that affect springer spaniels. Other fierce breeds also have low levels. High serotonin may hence be a key to the typical dog's calm personality. Drugs that influence its levels in humans also reduce aggression in dogs—a further hint that serotonin was a key to a place around the fire. Another receptor involved in nerve transmission is coded for by a gene which has a variant in humans said to be associated with curiosity and the search for new excitements. The variant is common among wolves, less so among aggressive dogs, and even less so in the fireside pet.

A remarkable experiment in Russia has begun to disentangle the chemistry of calm. In the 1950s, Dmitry Belyaev became interested in the inheritance of social skills in animals and man (a risky pastime at the time, for genetics was under attack from Stalin). He began to work on silver foxes, a coat-colour mutant of the red fox that had first been bred in captivity on Prince Edward Island in Canada in the nineteenth century. Its elegant fur was a valuable commodity in icy Russia, but the caged creatures were so aggressive and fearful that they were almost impossible to control. After a few years he moved his experiment to Siberia, where it continues today.

At first Belyaev's stock was suspicious and agitated in the presence of a keeper, albeit less so than the animals captured in Canada. Each generation, he chose as parents those best able to withstand the sight of a human. The rules were strict: just one male in thirty and one female in ten were allowed to breed. Within a few generations he saw a great change. The creatures became calmer and friendlier. They wagged their tails and learned to bark. Soon they revelled in their keepers' company, and for a time were sold as pets to raise funds. Even their appearance shifted, with piebald coats, curly hair, floppy ears, and blue eyes. Like dogs (but not like foxes) they have sex all year round—and like dogs they are good at the hidden food test, which the unselected foxes fail. Thirty generations on, almost all the animals are tame.

The newly tranquil beasts were cubs that never grew up. Over the generations selection for lack of fear led to an increase in serotonin. Among the few other genes that changed were some involved in the synthesis of the red blood protein haemoglobin, which were less active in the tamed animals. That seems odd, but those proteins also help soak up some of the chemicals involved in the serotonin response to stress.

The real revolution in the life of the silver fox took place before the Siberian experiment began. Their ancestors in Canada, when first taken into captivity, had found it almost impossible to cope with humans. On Prince Edward Island, where the "silver" mutation was discovered, it took years before any fox could be persuaded to tolerate the presence of a human being, let alone reproduce. Only after that barrier was breached could artificial selection begin and the rest of the emotional agenda follow. Like dogs, farmed foxes—both the friendly and the aggressive stock—diverge from their wild ancestors in the activity of a whole series of brain genes, some of which alter hormone levels. However, there are almost no discernible differences in the activity of a sample of brain genes between Belyaev's newly serene silver foxes and their unselected (and agitated) relatives in their cages; proof that friendliness demands fewer mental adjustments than does the simple but challenging ability to cope with human company. Most farm animals show few signs of amiability towards man, but for them and their owners acceptance is quite enough. To do so rescued them from the wild. The simulacrum of comradeship, as seen in tamed foxes and domestic dogs, came from changes in other genes.

The first farmers modified their charges in body and mind, and their modern descendants do the same—but the new way of life altered the ancient tillers of the soil as well. The new domestics, both plant and animal, became agents of selection upon those who had tamed them. The past ten millennia have been an era of unprecedented change for humankind, for *Homo sapiens* has evolved fast since agriculture began.

Our own equivalent to the silver fox mutation, the blond, is a creature of the fields. Just one person in fifty, worldwide, has fair hair. Before Columbus confused matters, almost all of them lived within a thousand miles of Copenhagen. Their relatives the redheads had their headquarters in Scotland, Wales, and Ireland. The half-dozen or so genes responsible for pale hair, skin, and eyes have become common only in recent times. Farming is to blame.

Northwest Europe, before the development of modern grain varieties, was the only place on Earth where such crops could be cultivated north of a line that passes, more or less, through Birmingham in central England. Wheat, barley, rye, and the rest need warmth to grow. In the Middle East, where they began, the sun shines upon the fields and those who cultivate them. In contrast the British landscape is dreary for many months of the year, and the peasants till the fields in gloom. As they do, the Gulf Stream imports energy from the tropics and heats up the ground even at the end of winter, when seeds need warmth. In the five thousand years since the first crops arrived in those northern parts, their grain-based economy has depended on Neptune's help.

A cereal diet (even when transformed into a sickly drink) is all well and good when supplemented by other foods, but is risky when life is just one grain after another (which for our peasant ancestors it often was). Cereals are low on vitamins; vitamin D, the antirickets substance, most of all. A move out of Africa had earlier led to the evolution of white skin to help northern hunter-gatherers to make the vitamin in sunlight. The arrival of farming in southern England marked a new challenge. In that earthly paradise the winter and springtime sun is so weak that a typical Greek, Spaniard, or Italian with dark skin and hair cannot make enough of the vitamin to stay healthy. Any child born in that dank climate to a dreary cereal diet and lucky enough to inherit a new mutation for fair hair and skin is at a real advantage, for the sun can then penetrate further into the flesh. The infant can make more of the crucial chemical for more months of the year and is safe from rickets and the other diseases that emerge from a shortage of vitamin D. The Age of the Blond (and the redhead) began with the first northern harvests. Quite soon, the homeland of the flaxen-haired expanded to overlap that of the boreal cereal growers, with its high point in Scandinavia, where more than half the population is blond.

I once spent a decade in Edinburgh and saw the sun for a few days. My present home in London has the equivalent of an extra month of full sunshine each year. Scotland has the worst health in western Europe, and vitamin D deficiency is twice as frequent as in England. Glasgow, its cloudiest city, has levels of chronic illness higher than any other British city. Climate may be as much to blame as its much-discussed fondness for alcohol, tobacco, and deep-fried Mars bars. Scotland has plenty of blonds, and the incidence of the gene for red hair, with the almost translucent skin that often goes with it, is almost one in three. Perhaps even that is not enough to generate the vitamins needed for health. The Scottish authorities now plan to provide vitamin D supplements to all those who need it.

Natural selection by diet acted upon the peasants in other ways. A muesli eater digests a lot of his breakfast before he swallows it, for enzymes in the spit break down starch into sugars that can be absorbed. People from places with a high-starch diet (northern Europe included) have up to fifteen copies of a gene for the crucial enzyme, compared with just four or five in peoples who eat mostly wild fruit, which contains sugar rather than starch. The early farmers found another marvellous way to consume grain (and relieve winter gloom) when they invented beer. That was bad for their brains but good for their guts, for bacteria do not like alcohol and ale was a safer drink than was the polluted water of ancient villages. Since brewing began, natural selection has done well, for almost all Anglo-Saxons can swill the stuff down. Most Asians cannot, for they lack the bibulous West's new and potent version of the enzyme that breaks down the poison.

Animals, too, changed the fate of their keepers. Most people across the world (and most adult mammals) cannot digest milk once they have left their mother's breast because they lack the requisite enzyme. It works in the small intestine to break an indigestible milk sugar, lactose, into two simpler sugars, each of which can then be absorbed. In many animals—and most humans—the gene is switched off soon after birth. An adult who drinks milk feels bloated or suffers from diarrhoea (fortunately, butter, yoghurt, and cheese do not generate that effect). For many northern Europeans, though, milk stays nutritious throughout life. A mutation that turned up soon after cattle were tamed allows the lactose-cutting enzyme to persist, enabling those who have it to digest milk when they grow up. Nineteen out of twenty Swedes but no more than one in ten Sicilians have that talent, and the map of its distribution fits with that of genetic diversity in the local cattle breeds (a hint as to how long they have been on farms). Eight-thousand-year-old remains of fat from cheese or yoghurt caked onto pottery fragments from Anatolia indicate that cows were being milked there at that time. Even so, the locals did not drink raw milk, for DNA in the bones of Europeans from three thousand years later still shows no signs of the lactose tolerance variant.

Long stretches of homogeneous DNA on either side of the genes for blond hair and milk digestion show that the new variants dragged their neighbours along as they swept through the population. Such stretches of homogeneity always hint at natural selection in action—although in most cases we have no idea what agent was involved or why. In time, the segments are broken up by the reshuffling that accompanies sex. A search through the DNA of people from Africa, Asia, and Europe reveals many such segments, each a relic of a sudden attack of selection in the fairly recent past. The change in lifestyle and diet that accompanied agriculture has caused evolution to speed up greatly compared with its average rate since humans split from chimpanzees. Man, like his animals, has seen many changes since he moved to the farm.

Domesticity itself arose long before humans began to till the soil, for even Neanderthals had a society far more sophisticated than that of any other primate. Agriculture moved it along. The notion of *Homo sapiens* as a house-trained ape is an old one. Darwin saw the parallels between the farmyard and the parlour: "We might, therefore, expect that civilised men, who in one sense are highly domesticated, would be more prolific than wild men. . . . The *increased fertility* of civilised nations would become, as with our domestic animals, an inherited character." He had hoped to add a whole chapter on humans to his work on the origin of farm animals, but he saw that the book was already "horridly, disgustingly, big" and abandoned the idea. That chapter has now been written by science.

As in the famous Siberian foxes, the real revolution in our own behaviour and anatomy took place when an ape became human. The hairlessness, upright gait, and agile mind that marked the transition were followed by physical changes that resemble those found in domestic animals. Compared with our ancestors, we have a lighter build, thinner skulls, shorter jaws, and smaller teeth, and the physical differences between males and females are less marked than before. We quarrel less about sex and are less enthused by it than are our primate kin and, like dogs, men and women copulate all year round rather than in a short season. Our breasts—like the cow's udders—are larger and milkier than theirs. Like pigs, we store fat more readily than do our untamed relatives, and we are less keen on physical activity. As in dogs, sheep, and cattle, various odd physical mutations (light skin and blue eyes included) have emerged in some populations, although humans have not yet gained a patchy coat. Unlike them, though, our brains have not become smaller.

Such changes in bodily structure came at a price: death or reproductive failure for the millions who could not cope. The speed at which advantageous genes have spread since the origin of farming suggests that the price was high.

The same is happening today. We now face not shortage, but a new abundance, different from anything in the evolutionary past, and have not yet evolved to deal with it. We may do so in time, but the process will not be cheap. Humans long ago evolved mechanisms to guard against excessive weight loss when food is short. Our bodies deal much less well with today's glut. Today's starvation disguised as surfeit means that Darwin's inexorable machine is cranking up again, and natural selection by diet may become as active as it was ten thousand years ago.

Anyone's response to excess depends in part on DNA. A study of twin boys and girls suggests that around seven-tenths of the variation in body weight within a population is due to genetic variation. Identical twin children also resemble each other closely in how much they will eat, given the chance. Adult twins paid to gorge themselves, or to starve, for several weeks also tend to gain or lose weight to the same extent, proof that their bodies resist, or surrender to, the challenges of the new diet to an extent that depends on DNA.

Hundreds of genes have been blamed for the new wave of obesity. Fatness certainly runs in families, but so do frying pans—and fat cat owners tend to have fat cats. The pets share their diet but not their DNA. Nature and nurture work together, and the inheritance of pot bellies is—like inheritance of almost everything else—not simple. The notion that fat people can blame their genes and ignore what they choose to eat is wrong: like alcoholics, they are more at risk of a certain kind of diet than are others and must struggle harder against temptation. Many people corpulent today would have been thin a century ago, whatever the structure of their double helix.

Even so, genes have a real influence on waistline. Most of those that predispose to obesity have a small effect, but one among them, when inherited in double dose (as it is by fifty million Americans), increases body mass by three kilograms above the rest of the population. Even a single copy, as borne by almost half the nation's inhabitants, adds a kilogram. The variant involved is active in parts of the brain involved in hunger and satiety. It is harder at work in starved individuals than in those who have just eaten. Although it does not affect weight at birth, babies with two copies begin to pile on the pounds in just two weeks. Other genes that dispose to obesity alter the efficiency with which food is soaked up or the rate at which the body burns its fuel.

The environment itself also has effects that stretch over generations. Just as alcoholic mothers have unhealthy babies, women who eat to excess have fat children, not just because they pass on their genes but because they were overweight while pregnant. About a third of all pregnant Americans—and half of those from the African-American community—are obese. Their internal economy shifts to deal with the problem and so does that of their developing child. The fetuses of obese mothers respond to the high levels of insulin (the hormone that controls blood sugar) in their mother's bloodstream and are born attuned to lay down fat. Genes for susceptibility to the diseases of the obese are then put under further pressure. Mothers who lose weight between one child and the next tend to have lighter babies than otherwise they would.

The biggest threat to the overweight is diabetes—not the rare variety that affects a few infants and is treated with insulin but a related illness that comes on later, defies treatment, and is strongly associated with an expanded waistline. The problem arises from a resistance to the hormone. Symptoms include heart disease, kidney failure, blindness, nerve damage, and even gangrene. Once a disease of the elderly, late-onset diabetes is now seen more and more in children and adolescents. Two extra inches on the waist add noticeably to the danger, and the risk to those in the top tenth of the weight range is twelve times that of the risk to the slim.

Half a billion people will soon suffer from the illness. Unless things improve, an infant born in the United States today has a one in three chance of developing the condition when it grows up. The cost of treatment already takes up a sixth of the country's entire health budget. In some places the figures are dreadful: half the adult population of the Pacific island of Nauru is afflicted. Genes that increase body weight when excess poor-quality food is available are much to blame.

Obesity is a target of natural selection because it kills people before their time. Its effects on the evolutionary future are made worse because the fat face not just premature death but sexual failure. Fat people tend to have fewer children than average. Apart from their romantic problems, obese men find it hard to sustain an erection, and overweight couples copulate less often than do the fashionably slim. Worse, an obese man's sperm count drops by around a quarter, perhaps because his overinsulated testicles are too warm. Female fertility, too, drops with every extra pound. Excess fat interferes with the menstrual cycle and has other harmful effects. Among women anxious to become pregnant, even a slight weight excess increases the delay before they succeed by almost a year. In addition, such women are more liable to miscarry, and their children are at higher risk of birth defects.

All this means that natural selection by diet is once more hard at work. Darwinians, faced with the problems emerging from the new agriculture, can hence afford a certain grim optimism about the future. Man evolved to deal with a new diet in the first food revolution, and will no doubt do so in the second, however high the cost. Even in an era of global glut, natural selection may in time act to return humankind to slimness and health, just as the descendants of the first farmers evolved their way out of their dietary problems.

The crude tools of evolution are, needless to say, far less effective in moulding the future than is mankind's simple ability to learn from its mistakes. Societies facing the waistline problem are better advised to eat less rather than await the attentions of Darwinism. Everywhere, people are exhorted to change their diet and to take up exercise (although so far the propaganda has not been very effective). Even Marie Antoinette was trying to help. The famous "cakes" offered to her starving people were not lard-laden delicacies but baked crusts that might otherwise have been thrown away. A simple misunderstanding gave rise to a legend of callousness and a sticky end. Her regal counsel now seems more sensible than it did at the time of the French Revolution. Will people take her advice and modify their lethal diet, or will they wait for natural selection to do the job? As Zhou Enlai said when asked about the success of the political event that killed the queen, it is too early to say.

CHAPTER 6

The Thinking Plant

DEEP IN THE AMAZON JUNGLE, A CREATURE SNAKES INTO THE light. As it climbs cautiously through the branches it senses a brighter spot on a distant tree. After weighing up the risks of abandoning its present post it plunges back into the gloom of the forest floor and creeps across the ground until at last it reaches its target, scrambles upwards, and triumphs to bask high in the tropical sunshine. The vine—for such it is—shows every sign of foresight in its behaviour. The notion that a plant might act in what appears to be an intelligent way seems alien; less so, perhaps, than before time-lapse films speeded up the circling of shoots or the opening of flowers, but at least unexpected. Can such a simple creature really plan ahead?

Romantics have long been convinced that the vegetable kingdom has a mind of its own. Gardeners talk to their crops in the hope that they will flourish, while real enthusiasts for botanical intelligence believe that cacti grow fewer spines when exposed to soft music and put them out again when a cat draws near. The Japanese even enter into two-way conversations with their green friends. They have patented an electronic device through which a flower can chat with its owner (or, when thirsty, ask for water). In the 1920s the great Indian physicist Chandra Bose, a pioneer in the study of electromagnetic waves, worked on electrical activity in plants. His subjects generated a measurable current when damaged (an observation that led to genuine scientific advances)—but Bose was also certain that music and kind words could set off the response.

Dubious as such claims might be, the mental universe of plants is, if nothing else, useful fuel for metaphor. Shelley writes of a garden in which a mimosa droops in response to a rejected lover's despair: "Whether the sensitive Plant, or that / Which within its boughs like a Spirit sat, / Ere its outward form had known decay, / Now felt this change, I cannot say." The Latin name for Shelley's sympathetic subject is *Mimosa pudica,* in reference to its bashful nature (the Chinese call it "shyness grass"). Whatever the plant's mental state, it does respond to the outside world. Most of the time, a mimosa's branched leaf stands proud; but a slight touch, or a gust of wind, causes it to droop. It can take hours to recover. At night, no doubt exhausted by the emotional turmoil of the day, the leaves close up and their owner goes to sleep.

Shelley's lines are both a literary device and an accurate observation. They also say something about the relationship of mind and brain. If a mimosa can act in an apparently rational way without any hint of cerebral matter, what does the endless debate on that topic mean? Philosophers, like poets, should pay more attention to botany.

Charles Darwin had no real interest in such metaphysical ideas (although he did claim that plants sometimes recoiled in "disgust"). He was nevertheless curious about plants' ability to react to the conditions in which they are placed, and wrote two books on the subject. *The Movements and Habits of Climbing Plants* of 1875 deals with how ivy, brambles, and the like find and scramble up their vertical helpers. *The Power of Movement in Plants,* published five years later, asks wider and more radical questions about how plants respond to the outside world. It had, he wrote, always pleased him "to exalt members of the botanical world in the scale of organised beings," and in those volumes he succeeded. Together they discuss three hundred species and place the plant kingdom on a higher scientific plane than ever before. His experiments laid the foundations of modern experimental botany.

The great naturalist's home county was in those days famous for hops. So fond was the British working man of beer that Kentish fields were filled with poles and wires up which the bitter vines were trained. Each September, tens of thousands of labourers and their families came from London to pick the crop and have what, in Victoria's glorious days, passed as a holiday. *Climbing Plants* asks a simple question. How does a hop find a support and climb up it? Its shoots as they peep above the soil must seek out an upright of the right size, even if it is a foot or more away. Then they must twine around it to clamber upwards. Such talents were the introduction to a new world of botanical behaviour.

Most of the research was done with the help of his son Francis. It was, as ever, interrupted by ill health: "The only approach to work which I can do is to look at tendrils & climbers, this does not distress my weakened Brain." Charles noted, first, that a potted plant in his sick-room circled round as it grew. He and Francis began to cultivate a variety of species beneath clear glass plates upon which the position of the tip could be marked with ink. They saw that the shoot of a young hop travels round all points of the compass. On a hot day a complete revolution took about two hours. Should the questing tip touch a pole, the hopeful climber changed its behaviour, snaked around it, and found its way to the top. What seemed like forethought depended on just three simple talents: the ability to circle, a sense of touch, and the capacity to tell up from down.

Father and son went on to study other kinds that climb not just with their shoots, but with structures such as tendrils, hooks, and adhesive roots. Whatever the details, almost all climbers gyrated until they found a support and, once found, clambered away from the ground. They found that all shoots, even those that do not climb, circle to some degree. In the same way, all plants can modify their growth to avoid an obstacle, and all can sense gravity. A hop's unusual powers depend—like so many patterns of animal behaviour—on natural selection's ability to modify existing talents.

The second book, *Movement in Plants,* goes further. It describes experiments on the sensitivity of roots, shoots, and other parts to light, gravity, heat, moisture, chemicals, touch, and damage. The research was far ahead of its time. Although they did not invent the name, the Darwins saw the first hint of the existence of hormones—not in animals (in which they were not discovered for another thirty years, when scientists at University College London found a chemical messenger in the blood of dogs) but in plants. So impressed was Charles by the powers of shoots and radicles (the first structures to emerge from the seed at the time of germination) that his book ends with a dramatic claim: "It is hardly an exaggeration to say that the tip of the radicle thus endowed, and having the power of directing the movements of the adjoining parts, acts like the brain of one of the lower animals; the brain being seated within the anterior end of the body, receiving impressions from the sense-organs, and directing the several movements."

Any creature, animal or vegetable, needs as it copes with the outside world to find out what is going on, to pass the information to the appropriate place, and to respond to the challenge. Men and women do the job with eyes, ears, nerves, brains, and muscles. Plants have none of those but manage remarkably well—and in some ways put our own abilities to shame.

Why climb? Lord Chesterfield got it right. In one of his notorious letters of advice to his son he wrote that "a young man, be his merit what it will, can never raise himself; but must, like the ivy round the oak, twine himself round some man of great power and interest." A plant with such an ambition uses its support to reach a lofty place to which it could otherwise never aspire. The helper might come to regret its generosity, but the advantages from the social climber's viewpoint are clear.

Such behaviour opens up a universe of opportunity. The earliest plants with flowers able to attract pollinators, and those with fruits that could persuade animals to eat and to move their seeds, discovered a whole new set of habitats and lifestyles. Their descendants were enabled to burst into a variety of forms. The ability to climb is less dramatic, but those who take it up have also evolved into a great diversity of kinds. A hundred and thirty different families in the botanical world have climbers. Within each family, those agile creatures are represented by many more species than are their earthbound kin.

Birds, bats, flying squirrels, snakes, and fish all take to the air, but in different ways, with modified arms, hands, bodies, or fins. In the same way, plants have modified a variety of organs to help them climb. Some, like hops or peas, use tendrils, based on stems or leaves. Others, such as clematis, have altered leaves in other ways or have evolved specialised roots or hooks that allow them to scramble. Roses have hooks. The ivy uses roots to clamber a hundred feet and more up cliffs, houses, or trees, while Virginia creepers go to the opposite extreme and use shoots. In a certain group of ferns the fronds grow around the support to make their way towards the light.

The habit is ancient. Three hundred million years ago, Earth had vast swamps filled with treelike ferns fifty feet high, later to be compressed into coal. The forest had plenty of vines and climbers, which used structures like those of modern plants to struggle into the sun. It became a tangled and impenetrable mass until at last the whole lot was wiped out as the climate changed.

Tropical forests are still the scramblers' capital. There, every plant must fight against thousands of others to reach sunshine. Many jungles are filled with lianas, woody vines that loop down from the trees. In most places they represent less than a tenth of the total mass of plant material—but their tactics are so effective that their leaves fill half the canopy. Almost half of all woody species in the Amazon basin are climbers, with fifty or more different kinds in every hectare. They are fond of gaps, places left open when a moribund giant crashes to the ground or when farmers cut a clearing (which is bad news for the farmers, who must compete with the creepers to grow a crop). When forests—tropical or temperate—are broken up by loggers, the lianas and their relatives thrive even as the trees upon which they depend are destroyed.

Climbers climb, in the main, to get into the light. Another good reason to grow away from the ground is to escape, like a baboon pursued by a lion, from predators. Leaves near the surface get chewed more by slugs, snails, and the like than do those up in the air. In the arid deserts of northern Chile, convolvuli often grow near cacti or thorny shrubs. After an attack by hungry mice, or by scientists with scissors, they at once increase the rate at which they twine and put out more tendrils in the hope of reaching a shrub and clambering into the safety of its spiny branches.

Darwin noticed that climbing plants needed a rather slender pole if they were to make progress and that British climbers never curl around trees. Their support must also be rough enough to give them a chance to hold on. The climber does not cling with its whole length but sets up a series of contact points as it moves on. Rather like a bloodhound it sniffs the air now and again as it tracks its route. As it moves, the tip is raised, circles round, and comes back to the stem a few inches further on. Some tendrils are set like a coiled spring to twist within seconds around a support as soon as they touch one. Engineers have worked out that a climber cannot ascend a smooth support more than about three times its own thickness—a twig, or a vertical wire. Much of a questing shoot's spiral motion comes from an increase in the rate of growth on one side of the plant compared with the other. In addition, cell walls on one side become looser, bulge up, and force the plant to wind round and round. In time, a tendril can coil in upon itself and grow hard and woody, to lock its supporter in a fierce embrace.

In certain species young stems are rigid and grow upright without help for several metres—but once they touch a tree, they pounce. No longer do they need to invest in solid—and expensive—wood. Instead they become thin and lithe and begin to clamber. Certain lianas grow a flexible stem to find the open air, but once they reach a sunny spot they generate heavy trunks that swing downwards from the heights and find another plant to use as a support.

The North American poison oak grows as a solid six-foot shrub when it stands alone, but extends ten times higher if it can find an upright. Many other kinds take advantage of a helper when they get a chance, but stand on their own trunks if they do not. In a tropical forest, young trees of many species not often thought of as climbers grow slim and tall by leaning against their neighbours. If that choice is not available, they too stand alone.

In many kinds, some branches have small leaves and move in a wide circle in search of a gap through which they can insinuate themselves. Those that sneak through and find the sun then grow larger leaves that soak up its energy. As the stems spiral away from the ground, they develop wide vessels through which to suck up water and food. In time, the liquid has to travel through a long passage, which makes drinking expensive and forces the plant to reduce water loss with waxy leaves and impermeable stems.

A tree pays a high price for its fellow travellers for they suck water and minerals from the soil and shade their host from the sun. West African trees in the presence of lianas grow at only a fifth of the rate of their fellows. A few plants of that kind can kill. The strangler fig, once it has reached the canopy, sends roots down from its eyrie. As they grow, the aerial roots wrap themselves around the supporter's trunk, fuse together, and squeeze it to death. The lethal tenant is left vertical and proud, with its own roots in unencumbered soil. In other trees the benefactor crashes to the ground under the weight of its visitor, but by then the fellow traveller has moved on in the canopy to bask in sunlight at another tree's expense.

Some plants twine clockwise and some counterclockwise—as in the famous case of the right-handed honeysuckle and left-handed bindweed. A mutation called "lefty" in a small mustard plant persuades the normally straight stem to spiral to the left, while another causes the opposite bias. Each changes the shape of a crucial protein in the cell skeleton. The molecule looks like a string of asymmetric dumbbells, with each element lying together head to tail to form a helical and hollow cylinder. The mutations enlarge one or other end of the protein and deform the cylinder, which alters the pattern of cell division and causes its owner to twist in one direction or the other. In an echo of the Flanders and Swann song, plants with a single copy of each mutation do indeed grow straight up (although they do not fall flat on their face). For reasons unknown, a bean that normally circles to the right much increases its yield if forced to twist to the left.

Climbing plants are of interest to gardeners, brewers, and wine drinkers, but to Charles Darwin they meant much more, for they introduced him to a whole new range of botanical talents. *Movement in Plants* shows that leaves, root tips, shoots, and other parts of all species, climbers or not, are in constant motion. They respond to circumstances just as animals do. Plants are slower, but they get there in the end.

The hop's abilities are eclipsed by the skills of all seedlings as they emerge into a hostile world to fight for light, water, or food. *Movement in Plants* contains a graphic description of the apparently purposive actions made by a newborn plant in its first days. In the struggle to turn into the right position, to push its root into the soil and its shoot into the air, a seed as it germinated reminded Darwin of a man thrown on his hands and knees and to one side by a falling load of hay. "He would first endeavour to get his arched back upright, wriggling at the same time in all directions to free himself a little from the surrounding pressure . . . still wriggling, would then raise his arched back as high as he could. As soon as the man felt himself at all free, he would raise the upper part of his body, whilst still on his knees and still wriggling."

To escape to safety the shoot must respond to light, gravity, physical contact, and other stimuli. Men and women live in a universe of senses—sight, sound, smell, taste, touch, and, the forgotten sense, position. Seedlings have no noses, tongues, fingers, or ears, but they too perceive all those attributes of the outside world. Animals use electricity and chemicals to pass messages through the body, and so do plants. Vegetation has no muscles—but its bearers grow to where they need to be, or move with the help of molecular machinery quite like that which drives our own limbs. As *Movement in Plants* put it, "It is impossible not to be struck with the resemblance between the . . . movements of plants and many of the actions performed unconsciously by the lower animals . . . the most striking resemblance is the localisation of their sensitiveness, and the transmission of an influence from the excited part to another which consequently moves."

Without eyes, ears, or nerves, how can a plant know which way is up, what has touched it, or whether the sky is blue or grey? We are beginning to find out.

Father and son identified two kinds of activity—those in which only the response, and not its direction, is related to the external trigger and those that involve a move towards or away from an outside stimulus. Among the former, they noted that many plants opened and closed their flowers in sunlight, or "went to sleep" by folding their leaves at night, perhaps to reduce heat loss by radiation. Some, like the mimosa, responded to a sudden prod with a collapse of the leaves in an attempt to frighten off a hungry insect, or to expose an enemy to the thorns on its branches. All those with sensitive leaves slept at night, but plenty of sleepers were indifferent to a poke.

In the mimosa and its fellows, the movement comes from a sudden loss of internal pressure in each leaflet of the frond, which spreads to those next to the one touched. Certain cells held in a bulge at the base of the leaf stalk are crucial. If they are tickled the leaf folds at once. They also control the response to darkness or light. The hinges are more sensitive than human fingers. Each has a long hair that acts as a lever and is embedded in a sensory cell. On a stimulus, the movement at the base of the hair sets off a response as ions are pumped across the cell membrane. At once water is lost, parts of the internal skeleton collapse, and the leaf folds up. In time, the plant forces water back into the hinge, which reopens, ready to respond to the next challenge. The pattern of two opposed forces—one that collapses and one that opens the leaf—is rather like our own arrangements, in which one muscle causes a limb to extend and another makes it flex.

Many flowers can tell the time, and the ancients set the hour of prayer with a quick glance at the garden. Linnaeus, the great classifier of life, designed a bed filled with blossoms that opened at different hours to serve as a crude botanical timepiece. Many plants—such as sunflowers, which track the sun through the day—respond only to outside stimuli. Mimosas have a more subtle sense of the hours, for when placed in constant darkness the rhythm of sleep and wakefulness persists. They have a biological clock that runs on what Darwin called "innate or constitutional changes, independent of any external agency."

An internal timer, based on the buildup and breakdown of chemicals, maintains the rhythm. The clock is not precise and will wander away from time if kept in constant light or dark. Different species have internal timers with a cycle that varies from around twenty to thirty hours. Dawn resets the mechanism, which hence keeps up with the changing hours of daylight as the seasons progress. The inner and outer worlds interact, for mimosa leaves do not fold up at night unless they have been illuminated during the day.

Such movements may look purposeful, but they lack direction. Other botanical talents seem, in contrast, to have a definite goal in mind. A plant's life is ruled by sunlight, water, food, and predators. To survive it must avoid its enemies, find its friends, and, like an animal, find food, water, shelter, and—most of all—the light of day.

The Darwins found that young shoots will grow towards even a dim source of light. That simple observation led them to their most significant result: the discovery of an internal messenger—what we would now call a hormone—that altered growth. It was the first of thousands of such chemicals now known.

Their experiment was simple but ingenious. A shoot of grass bent towards the light. It did so, they found, only if the beam hit its topmost point from one side. Crucially, if the very tip was covered, or if the ray was directed to a spot just below it, the shoot remained unmoved. When the plant was buried in sand with only the tip left in daylight and the rest in blackness, the buried shoot grew towards a lamp even though its rays never touched the growing part. Short bursts of illumination had the same effect as a single longer glow, and even low levels did the job. The very tip of the shoot, they realised, was sensitive to light, and somehow the information on where it came from was passed ("influence is transported") to the stem below. In turn that persuaded it to alter its activity.

Years later, in 1913, came direct proof of a chemical messenger. The amputated tip of a stem was placed in daylight on a piece of sponge. The sponge soaked up the crucial substance as the scrap of tissue pumped it out, and when it was laid onto a cut stem whose own tip had been removed, the shoot altered its pattern of growth. The botanical envoy was named "auxin" (after the Greek *auxein*, "to grow"). It was the first plant hormone.

For plants and animals alike, to learn about the world outside is not enough. For the organism to respond to the messages of opportunity or danger that pour in, information must be transmitted to a structure that can act upon it. It travels through an animal's body through many routes. Nerves pass it from place to place (and all cells, nerves or not, talk to their neighbours in one way or another), and tissues communicate with the help of hormones or other messengers released into the bloodstream. Plants have no nerves or bloodstream, but they too pass instructions through special channels that traverse the thick cell walls and allow their living parts to touch. The light-induced messenger is passed downwards from the shoot tip in that way and such channels also transfer proteins, nucleic acids, and a variety of hormones. Other signals tell the dark world beneath the soil when spring has come while yet more regulate growth or warn of the arrival of a predator. Plants also have open vessels but, unlike blood, liquid does not circulate but moves in a single direction, from roots towards shoots or leaves, from where it is lost by evaporation. The vessels transmit molecules that control cell division and cell death and others that say when it is time to flower or to store food.

Plant hormones resound through tissues in response to light, heat, damage, the passage of time, and more. Some turn up in unexpected places. Human urine applied to a decapitated shoot alters the shoot's growth because auxin passes unchanged through the body of those who eat plants (and the substance was itself first purified from that invaluable fluid). Now the messengers are studied not just with chemistry but with mutant plants whose altered growth is due to an aberrant response to a chemical message.

Most plant hormones are simpler and smaller than our own. Some are based on closed carbon rings while others are small proteins and few look rather like the steroids that control human sexual attributes. Like our own internal envoys, they are arranged in pairs, with some that promote an action and others that oppose it. Each has a receptor on the target tissue to which it binds, and each acts to stimulate or repress the action of particular genes. Some cause cells to expand or contract, while others change the rate of cell division—for example, causing cells to divide faster on the dark rather than the light side of a shoot, so that the structure bends towards the source.

Such molecules determine when a plant will ripen, shed its leaves, move towards or away from light and gravity, fight infection, and more. Those in the tip of a shoot suppress the activity of sections of the plant that lie below them. Auxin moves downwards and prevents the growth of buds that might compete with the tip for light. Cut off the tip, and those segments burst into life—which is why gardeners prune fruit trees to get a dense bush.

The locks into which the auxin keys fit have been discovered. Many of the changes in shape treasured by gardeners result from errors in the hormone genes or their receptors. Auxins persuade the shoot to grow up or the root down, the flower to bloom, and the fruit to swell. The sinister movement of the sundew as it rolls its leaves over a trapped insect is due to another member of the same family. In their movements from cell to cell the auxins resemble substances involved in nerve transmission as much as they do animal hormones. Indeed, some of our own nerve transmitters are found in plants, although what they do is not clear.

Auxins have been turned to useful ends. Gardeners and farmers use artificial versions to help cuttings to take root. The Vietnam War defoliant Agent Orange—named for the colour code on its barrels—was an artificial auxin that made vegetation grow itself to death. Twenty million gallons were used in an attempt to destroy the enemy's crops and to open up the forest to expose the guerillas. Its military value was never proved, and its use was abandoned when it was found to be contaminated with the poison dioxin. Even so, synthetic auxins are still used as herbicides and appear to be safe.

The leaves that hid the Viet Cong formed a dense screen as the trees that bore them struggled for life—and for light. All plants need sunlight and will fight to the death for it. Every forest is a silent battle among the leaves far above as they struggle for a place in the sun. Together, they can block its rays. Some bathe in its beams, but others fail, turn sickly, and die. To survive, the plant needs to detect solar radiation, measure its intensity, and move or grow in response.

Although they found that shoots can pick up light the Darwins had no idea how the plants noticed its presence, nor did they know that they were sensitive to its wavelength, intensity, and direction as well. A variety of special molecules, some of which have equivalents in the animal kingdom, help them do the job.

One group, the phototropins, picks up blue light and plays a large part in the growth of shoots towards a source of illumination (they also control the opening of the pores through which air and water travel). Others, the phytochromes, are sensitive to longer waves in the red and infrared. They have a protein skeleton matched with a structure based on rings of carbons bent into a molecular knot. The protein is poised like a set mousetrap, and when light strikes the molecule it flips from one shape to another. In dark or shade the change is slowly reversed. The balance between the two is a measure of how much light is in the sky.

In a dense forest, the amount of infrared radiation reaching the ground is no more than a twentieth of that experienced by a plant exposed to full sun, for the rest is soaked up by the green leaves above. Because phytochromes measure not just light intensity (which changes through the day) but also the ratio of red to infrared, they can distinguish a shortage of light in the evening or under clouds (when the proportion of the two wavelengths does not change) from a scarcity that arises because other leaves have shaded out the sun. Once the infrared alarm has sounded, those in the shade must respond before the energy source is blocked altogether.

A plant in this predicament shifts its whole pattern of life. It grows faster and the stalk stretches higher. Each leaf moves to present a flatter surface to obtain maximum exposure to radiation. If the problem continues, the leaves become thinner and more transparent and the plant becomes less branched as it reaches for the sky. If all this fails and the light still stays red, the unpalatable truth becomes clear. The victim flowers as soon as it can in an effort to pass on its genes before it dies in darkness.

The phytochromes are smart, but other pigments are even smarter. A third set of sensors, the cryptochromes, respond not to red but to blue light (in a parallel to our own eyes, with their three distinct receptors, a green-sensitive pigment has also been found). Cryptochromes have a structure rather like that of the enzymes that cut and splice DNA. They measure intensity rather than wavelength. Their main job is to sense how long each day is—an important factor when deciding whether to make flowers or fruit as the season moves on. In a newborn seedling wriggling its way towards adulthood, they shift the whole biochemical economy from an existence based on the gloomy world of the soil to a life bathed in sunshine.

Our own eyes are blessed with relatives of the plant sensor molecules. They too work out the length of each day and assess wavelength (or colour, as we call it). Some are the very stuff as dreams are made on. They not only cause a sensitive plant to droop but also control human sleep rhythms. A long plane flight leads to unpleasant side effects—and the blue-light pigments put them right (which is why a global traveller finds it harder to adjust to local time in dreary London than in sunny Sydney). Mice in which the blue-light gene has been damaged by mutation sleep more and have less active brains as they snooze. Its level in our own eyes are tied to the swings of mood familiar to those with seasonal affective disorder, the Black Dog of winter. Sensitive plants droop in grey weather with the help of cryptochromes, and so, it seems, do we.

Roots, in contrast to ourselves, try to avoid the sun. Again, blue light does the job, with its own special receptor molecule in its tip. The root has another talent that helps it delve into the soil, for it can sense gravity, as can, in the opposite direction, the plants that climb. As Darwin found, the crucial sense organ resides in the tip of the root and the shoot. Now it has been tracked down—and, once again, it has uncanny similarities to our own senses.

Men and women maintain their equilibrium with a set of liquid-filled tubes in the inner ear, arranged in three dimensions—left and right, forward and back, up and down. The tubes contain a liquid that washes back and forth as we stand, sit, or move about. Tiny grains of limestone rest on special cells on the inner surface of each tube and shift as gravity or acceleration moves them in one direction or another. The movements of the fine hairs on each cell are translated into electrical messages to the brain and give us a sense of where we stand.

Plants do the same in roots and shoots. Each contains small grains of starch, which, like the minute particles in our ears, shift as their owner moves.

Mutants unable to make starch lose both their sense of gravity and the ability to move in a circle. Darwin speculated that the questing movements he found in hops depend on the Earth's attraction, but he was not entirely right. In an experiment that would have flabbergasted him, the tips of plants held in weightless conditions on the space station continue to make their measured rounds.

A closer look at both people and plants shows further parallels. The ear has a molecular rack and pinion that uses a pair of proteins to pick up the movement of the grains as they are washed back and forth. In plants, the job is done by a pair of almost identical molecules.

Poets, mystics, and romantics often imagine the vibrations of a sixth, seventh, or eighth sense. One popular candidate is magnetism—a topic tarnished from its earliest days, when the German mountebank Franz Anton Mesmer claimed that "animal magnetism," the supposed ability of some people to open blocked bodily channels, could cure blindness and more (an idea blown out of the water by Benjamin Franklin). Magnetic therapists still sell hundreds of millions of dollars' worth of magic bracelets in the United States.

Biologists have renewed their interest in our interactions with the Earth's magnetic field. Some say that blindfolded students can find their way home thanks to a supposed internal compass, while aerial photographs hint that cattle tend to line themselves up facing north or south. A strong magnetic field does spark brain activity, but its relevance to daily life is not clear.

Many creatures have an unexpected ability to sense compass direction. Migratory birds have iron-rich cells within their skulls and use them to find their way across the globe as the seasons change. To do so, they use gene products remarkably similar to a plant's blue-light sensor. The molecule forms part of an internal timer with which the migrants use the angle made by the sun as it sweeps across the sky to orient themselves north or south. In addition, it helps the birds to navigate by the Earth's magnetic field.

They can pick that up only in daylight, and the blue sensor gives them the power to do so. It works at the atomic level. Electrons come in pairs that spin in opposite directions and, as a result, cancel out each other's ability to act as magnets. When energy—from light, heat, or chemistry—enters the system, it is disturbed and some particles are left with just a single spinning electron. Such "free radicals" are anxious to join up with another partner as soon as they can, by changing their direction of rotation. A magnetic field makes that harder. As a result, a bird can use the pattern of unpaired electrons as a compass needle. The energy needed comes from the blue light picked up by the cryptochromes.

Magnetism also affects plant growth. The starch grains in the tip of a shoot can be moved with a magnet, and the field helps tell a plant where it stands (an attempt to test that process without gravity perished in the *Challenger* space shuttle disaster). Mutants that lack cryptochromes are indifferent to its presence. Perhaps all this hints at a forgotten sixth sense shared by animals and plants. Darwin often spoke of "fool's experiments," ideas that were most unlikely to work but worth a try. Though he speculated on the role of magnetism in animal navigation, he would be amazed to find that it applies to the plant kingdom; the experiment was too foolish even for him.

In another manifestation of what Francis called his father's "wish to test the most improbable ideas," Charles persuaded his son to play the bassoon to a mimosa to test whether it would respond. Sweet music, like kind words, had no effect (in a modern experiment pop music blasted at them for hours also left them unmoved). Even so, plants do hold conversations, using not sound but scent. A series of chemical messengers provide a sense of smell to add to those of sight, gravity, and the rest.

Many plants pump out a gas called ethylene which causes them to grow faster and also helps fruit to ripen. In dense vegetation the gas, which reveals the presence of competitors, persuades leaves to struggle harder towards the light. The botanical ability to smell can be used for more sinister ends. Dodder is a pest of carrots, potatoes, clover, and garden flowers (it is also known as devil's guts, witch's shoelaces, and hellbine). Its leaves are tiny and contain almost no chlorophyll. The seeds can survive for ten years. Once they germinate, the shoot pokes above the soil and searches for a potential victim, circling round until it succeeds. Then it inserts fine needles whose cells fuse to its quarry. They suck out vital fluids and the dodder loses its own root, to live entirely as a parasite. It prefers juicy species like tomatoes over tougher kinds such as wheat.

Dodder, like a bloodhound, sniffs out its prey. When allowed to germinate in a closed container with two tunnels, one leading to a chamber with healthy tomatoes and the other to a similar chamber with wheat, it directs its growth towards its preferred host—the soft and juicy tomato—and is repelled by its alternative.

Some plants talk to themselves. A damaged leaf causes others nearby to get ready for attack—but not when it is sealed in a plastic bag, proof that an airborne signal is involved. Leaves chewed by insects pour out signals that are picked up by their neighbours, which prepare their defences. Other species may listen in. Tobacco seedlings grown close to a sagebrush switch on anti-insect chemicals when the sagebrush is pruned. A few plants can even call for help. A beetle that chews on the leaf of a lima bean is taking a risk, for its victim sends out an aerial message that persuades other leaves nearby to make a sugary secretion that attracts ants and wasps. These insects then attack the beetle.

Plants can also taste chemicals in solution. They extract information from the liquids that bathe their roots and flow across their leaves. Roots and shoots sense the presence of enemies and grow away from them. They hunt for food: when a root hits a rich spot, it stops, sprouts, and sucks up what is on offer. Material pumped into the soil by roots may attract friends such as helpful fungi, but is also hijacked by enemies. Witchweeds are pests of tropical crops such as sugarcane. They grow on the roots of their host and, like the dodder, suck out its vitality. They pick up the taste of a dissolved substance used by the host to attract fungi. In the same way, corn seedlings whose roots are chewed by grubs pump out a chemical that attracts predatory worms. The American black walnut scares off competitors with its secretions and leaves a dead zone beneath its shade—and it is no coincidence that our own tongues are titillated by the poisons found in pepper, coffee, lettuce, and more, which evolved not to satisfy gourmets but to fight off enemies.

Darwin saw that plants have a sense of touch, for climbers, as soon as they contact a vertical object, give up their wide sweeps, and begin to twine. Roots, too, probe the soil and grow around a stone too large to move. For roots, touch and the sense of direction interact, for a root held vertical will grow away from an object that blocks its path—but the same structure kept horizontal will always try to go beneath it.

Tree huggers carry out what seems the entirely witless experiment of embracing a trunk to exchange energies with it, injecting—or so they believe—their own vitality into the plant and obtaining some undiscovered botanical spirit in return. To touch—or to hug—a plant has another unexpected effect: it inhibits growth. Pines growing high in the Highlands of Scotland are small, twisted, and bent because they have been caressed—or battered—by the winds. Their equivalents around Down House live in calmer air and soar upwards. Identical seedlings grown in calm or windy places end up looking very different. To bend a young tomato plant for half a minute stops its growth for a whole hour, which is why squally places make for stunted trees.

The most conspicuous response to touch is that of the mimosa, so admired by Shelley. Two centuries after his poem, botanists returned with a simple experiment: take a series of chemicals known to act as hormones, dissolve them, and water the mimosa to see which bits of DNA respond. For every substance, a previously unknown gene increased its activity by a hundred times. At first it seemed that they had found a crossroads in the hormone labyrinth—but to sprinkle the plants with pure water had the same effect. Mere contact had done the job. The first of the "touch genes" had been discovered. A drop of rain or a gust of wind causes them to leap into action. More than five hundred separate parts of the genome increase its activity when a leaf is prodded, some by ten times and more. A hundred respond in the opposite way. Why, nobody knows.

The touch genes do many things. They cause the cell wall to firm up or loosen in response to stress, and growth patterns to alter as a result (which explains the gnarled Highland trees). Some members of the clan respond to night and day or to the seasons (which is why leaves fall off in autumn), and others are involved in disease resistance. They might some day be engineered to produce fruits that fall from the tree in a breeze as soon as they are ripe, or crops that grow tall in windy places. Half the touch genes also respond when placed in darkness, although why is not clear. The sensitive plant is, it appears, more sensitive than anyone imagined.

The inner world of plants turns out to be almost as rich as our own. Darwin was rightly cautious in drawing parallels between the two kingdoms' sensory lives, and the most he would say is that "it is impossible not to be struck with the resemblance between the foregoing movements of plants and many of the actions performed unconsciously by the lower animals." Now we know that the similarities go further than he thought. One telling correspondence involves our own sense of touch. Quite unexpectedly, some of the signal proteins that plants use to respond to a gentle tap resemble certain molecules that do a similar job for us. They control our heartbeats, switch on hormones that determine growth, and alter the blood chemicals that change moods from happy to depressed. In a further twist to the tale, young rats nuzzled by their mothers respond with an increase in the activity of certain genes related to those that react to touch in plants. A shortage of caresses stunts the animals' emotional growth.

There is something magical about the ability of scientific rationalism to connect raindrops with heartbeats and battered trees with depressed infants. Shelley saw that science told us much that poetry cannot. He filled his

Oxford room with electrical gadgetry and saw no contradiction between the worlds of the spirit and of science. He would have been delighted to learn that cooling passions are linked to falling leaves and that the Darwinian universal of shared ancestry shelters beneath its ample branches both the mimosa and its poet.

CHAPTER 7

A Perfect Fowl

SIR ROBERT MORAY WAS A SPY FOR CARDINAL RICHELIEU, A FREE-mason, a member of the Scottish army that took Newcastle from the English in 1640, and in his spare time the first president of the Royal Society. He wrote extensively on the natural history of his native land and made a remarkable discovery, published in the Society's *Philosophical Transactions* in 1677. On a log on the shores of the island of Uist, he saw "multitudes of little shells; having within them little birds perfectly shaped, supposed to be barnacles. . . . This bird . . . I found so curiously and completely formed, that there appears nothing wanting, as to the external parts, for making up a perfect Sea-Fowl; . . . the little bill like that of a goose, the eyes marked, the head, neck, breast, wings, tail and feet formed like those of other water fowl, to my best remembrance." Sir Robert had the honesty to admit that he had never observed any of the adult animals but told his readers that "credible persons have assured me that they have seen some as big as a fist."

The myth of the shell-born birds—barnacle geese as we call them today—the shells themselves supposed to be the seeds of a certain tree, was widespread. So embedded was the notion that for a time the barnacle goose was counted as a fish and could be eaten by Catholics on Fridays. Their breeding grounds are in the Arctic, which baffled those who saw them only in the winter (although Thomas Henry Huxley suggested that the mistake came because such birds were common in Hibernia, Ireland, and that the linguistic shift from *Hiberniculae* to *Barnaculae,* or barnacles, was easy enough).

The idea of a bird-bearing tree is foolish, but it arises from the ancient and accurate observation that the adult form of many creatures is quite distinct from that of their eggs or embryos. The untrained eye finds it hard to tell juveniles apart. A month-old human fetus is almost identical to that of a chimpanzee, the inside of a goose egg looks much like that of an ostrich, and a barnacle larva is not very different from those of its relatives among the lobsters and crabs. Even the founder of modern embryology, Karl von Baer, found it difficult to sort them out. In 1828 he wrote that "I have two embryos preserved in alcohol that I forgot to label. At present I am unable to determine the genus to which they belong. They may be lizards, small birds, or even mammals."

The word *evolution* was first applied to the unfolding of the body as egg is transformed into adult. Development is the imposition of pattern upon a formless mass. Most animals, from barnacles to geese, share the same basic types of cells. As the embryo grows they are organised to make a crab or a barnacle, a goose or an ostrich, a man or a bat. That grand reshuffling builds new and complicated body shapes from the same raw material. As it does, it hides the blueprint from which their bodies are built. *The Origin* used the similarity of the juvenile stages of apparently unrelated beings to argue that "community of embryonic structure reveals community of descent." Adult anatomy makes sense only when seen through the eyes of an embryo.

Darwin saw that many creatures showed "unity of type," a deep similarity hidden by the complexity of mature animals but manifest in their young. Many embryos—those of barnacles included—consist of repeated segments that multiply, reduce, or rearrange themselves to produce an adult. An increase or decrease in number or a shift in pattern of growth generates a vast diversity of sizes and shapes. Evolution works as much by manipulating repeated units as by tinkering with individual organs as they grow.

The idea finds new life in modern biology. DNA, like the bodies it builds, is a series of variations on a structural theme. Complex organs—eyes, ears, hands, and brains—are pieced together from units that can be seen only in the early days of development or, even earlier, revealed to be hidden in the recesses of the double helix.

Nowhere is the contrast between young and old more remarkable than among the barnacles. Such creatures were once thought to be snails because of their solid shells (a distinguished professor of zoology—or of biochemistry masquerading as that science—once tried to convince me that they do belong to that family). In fact they are arthropods, jointed-limb animals related to crabs, spiders, and flies. Their ancestors lived free in the oceans, but many now spend most of their lives in a prison. Barnacles are close kin not to limpets, as once thought, but to shrimp and lobsters. That affinity was discovered in the 1820s, but for many years the group—the cirripedes, or "curly-footed," to give them their technical name—seemed no more than obscure. Few biologists could be bothered with such tedious creatures.

Until, that is, Charles Darwin spent a sixth of his scientific career on them. His eight years of research, in the interval between the *Beagle* voyage and *The Origin of Species*, showed that such beings, dull as they might appear, had lessons not just for naturalists but for biology as a whole. As he came slowly to the idea that life was not fixed but might change, he was warned that "no one has the right to examine the question of species who has not minutely described many." Perfectionist as ever, he agreed: to understand the logic of life he needed to become an expert on a single group. Today's biologists are obsessed with "model organisms"—fruit flies, a certain worm, mice, mustard plants, and even humans—that might, when their secrets are unveiled, be exemplars of evolution on a wider stage. Cash pours in, and optimists hope that by understanding their favourites in detail the wider science of life will be illuminated. Some of the supposed archetypes turn out, alas, to be untypical even of the group to which they belong (the fruit fly falls into that category). The first model organisms of all were Darwin's barnacles. He made, by luck or judgment, an excellent choice.

He wrote four books—a total of one thousand, two hundred pages—on their taxonomy, their embryos, and their fossils. Some species were bizarre: so distinct from the familiar rock dwellers of Welsh or Scottish shores that any kinship with them seemed almost as improbable as an affinity with geese. The young naturalist did his job so well that, at the age of forty-four, he was given the Royal Society's Gold Medal for his research. His solid volumes remain a standard reference work. More important, they laid the foundation of a central theme of evolution: that the embryo is the key to the adult.

Darwin's attention was drawn to cirripedes when, as a medical student in Edinburgh, he spent weeks in the hunt for marine animals in the Firth of Forth. There he fell under the influence of the zoologist Robert Grant, who introduced him to life on the seashore and encouraged him to publish his first scientific paper (Grant later became professor of comparative anatomy at University College London, but he and Darwin fell out over the issue of whether life showed inevitable progress from low to high, and—although for a time they inhabited the same London street—scarcely spoke again). Almost a decade later, on the shores of the Chonos archipelago, off the coast of Chile, the *Beagle*'s naturalist found an enigmatic soft-bodied creature a tenth of an inch long drilling into a conch shell. At first he thought it was a worm, but under the lens it became clear that the animal was a great anomaly, for though naked it looked very like a British barnacle. Might it, despite its lack of a shell, be related to the denizens of a Scottish shore? If so, how—and why was the creature so different?

Darwin tried to find out. He planned at first just to sort out the Chilean's place in nature, but as the work went on he found more and more distinct and—on the face of it—aberrant kinds. Soon he began to notice what appeared to be intermediate forms, and series that showed greater or lesser affinity to each other. Although he was oppressed by the tedium of his task ("I may as well do it, as any one else"), barnacles sharpened in his mind the idea—already implanted, as his notebooks show—that one species might change into another. Perhaps, he concluded, all barnacles descended from a common ancestor. Indeed, what was true of barnacles might be true of the whole of life. Five years after his cirripede opus, that radical notion became the theme of *The Origin of Species*.

The juvenile stages revealed unexpected connections between the South American borer and its Scottish kin. That lesson, learned on the shores of Chile, has grown into the science of evolutionary developmental biology, which unites barnacles around the world with each other, with crabs and lobsters, and even with geese. It reveals the common foundations upon which each one is built.

Many creatures share a series of genes that lay down the basic body plan. They are control switches that guide the organism through its journey from fertilisation to the grave. Errors lead to dramatic shifts in form—legs growing in the place of eyes in fruit flies, lambs with two heads, or extra fingers in human babies—together with greater and more persistent changes such as those that made birds from dinosaurs or barnacles from the ancestors of crabs.

Darwin was sent specimens from around the globe. Some, he realised, would stretch the belief of his fellows; he wrote to a colleague about one of his discoveries that "you will think me a Baron Münchausen among naturalists." His first job was to describe what the creatures looked like. As ever, he told a straightforward story in plain prose.

The introductory paragraph of the first *Barnacles* volume is a sober account of what most people imagine such creatures to be: "Almost every one who has walked over a rocky shore knows that a barnacle or acorn-shell is an irregular cone, formed generally of six compartments, with an orifice at the top, closed by a neatly-fitted, moveable lid, or operculum. Within this shell the animal's body is lodged; and through a slit in the lid, it has the power of protruding six pairs of articulated cirri or legs, and of securing by their means any prey brought by the waters within their reach. The basis is firmly cemented to the surface of attachment."

That statement introduced the immense variety of cirripede lives. More than twelve hundred different kinds are now known, and no doubt many remain to be discovered. All live in salt water. They fall into two main groups, those with a stalk (the goose barnacles, a crab-flavoured delicacy in many parts of the world, named in homage to the ideas of Sir Robert Moray) and those without (many, like the familiar acorn barnacle, attached to rocks and other marine structures). Almost all have jointed legs, often tucked away within a shell. Most use them as a net to sweep the seas (although the stalked versions depend more on the movements of the water to bring food). Unlike their relatives the crabs and lobsters, barnacles do not moult their skeletons to grow. Instead their plates enlarge as individuals get older. Some sit on rocks, while other kinds burrow through solid stone or into snail shells or spend most of their time afloat. Yet others are parasites of crabs, jellyfish, and starfish. Some among that group are so specialised that, when adult, they look more like a fungus than an animal.

Like insects, barnacles have a head and thorax and, in a few species, what might be a vestigial abdomen. They have six pairs of hairy and jointed legs (fewer than the prawns and lobsters, which have ten). The familiar seaside versions spend their lives upside down, standing on their heads and waving their feet in the water in search for food.

Those found on rocky shores live in a fortress made of tough plates, based, like a snail shell, on a limestone-like substance. Different varieties have more or fewer segments of body armour. The average is six, and the great man devoted many pages of his books to the minutiae of how one might use the numbers of plates to sort out patterns of relationship. For the common North Atlantic form, an additional two plates act as a lid, which opens to let out the legs at high tide and closes to keep in water when the creatures are exposed to the air (which for some individuals is all the time, except for a few hours each month at spring tide). The mouth has structures that chew and grind and look a little like those of crabs and even of cockroaches. Some species excrete through their mouths, as their anus has faded away. The eyes, tucked away in the dark, are lost as well. The nervous system is reduced compared with that of the barnacles' free-living relatives, although they do have a heart.

However dull their cloistered existence within a gloomy fortress, barnacles have a remarkable sex life. Like all good biologists, Darwin spent much time on that topic. He found a wild diversity of reproductive habit. The textbooks of his day said that all were hermaphrodites, but many, he found, were not. Some have two sexes, some are male when young and female later, and some are true hermaphrodites—although a few among that group generate small males around their bisexual selves in case they might be useful. Many of those with conventional reproductive habits (and two sexes) spend their adult lives fixed to a single spot. As a result a male must constantly wave his penis (erected at the cost of lots of hydraulic energy) to reach out and tap his neighbours in the hope that at least one is a female. A male that finds itself in a sparse and scattered group must grow a longer organ than one living in a crowd. A female, once tapped, may copulate with half a dozen partners in series and then pump out most of their seminal fluid as not up to scratch. Her fertilised eggs develop into the first of several larval stages.

The young biologist's studies on cirripede sex brought forth some poetic paragraphs. The masculine organ of a certain species was "wonderfully developed . . . it must equal between eight and nine time the entire length of the animal! . . . there [it] lies coiled up, like a great worm . . . there is no mouth, no stomach, no thorax, no abdomen, and no appendages or limbs of any kind." In another, the males were reduced to parasites within the female: "thus fixed & half embedded in the flesh of their wives they pass their whole lives & can never move again." The creatures hold the record for relative penis size (a certain beetle comes next, at twice its body length). The organ comes at a price. Individuals from wave-battered shores have shorter and stouter members than do those from calmer places, as the male finds it hard to control a lengthy structure in a turbulent world. So expensive are their genitals that many males lose them at the end of each season and grow a new set the following year.

Barnacles are remarkable for reasons that stretch beyond the penis. They stick to their rocks with a sophisticated cement, a protein that repels water. It is the toughest known natural glue. Like an epoxy, the adhesive is secreted as a clear fluid with two components. When they mix, cross-links are made between the molecules, and the animal becomes almost impossible to dislodge. So powerful is the bond that some of the proteins involved may soon be used in surgery.

Barnacles cling to ships just as well as they do to rocks, and mariners have good reason to despise them. Darwin, exhausted by his years of work, wrote that "I hate a Barnacle as no man ever did before, not even a sailor in a slow-sailing ship." The *Beagle* herself had to have her bottom cleaned several times in the cruise around South America. A ship uses 40 percent more fuel when covered with marine organisms than when its surface is smooth—which is expensive and, in these days of the greenhouse effect, also to be deplored on ecological grounds. Poisonous paints were once used to keep bottoms clean, but most of them have been banned as pollutants. The best protection is to find a finish to which the animals cannot attach. Given that they can adhere to a nonstick saucepan, the job is not simple, although paints with added carbon nanotubes offer some hope. Some corals and seaweeds manage to stay free of encrusting creatures not with poisons but with chemicals that scare them off. Those unknown substances will make the fortune of the first scientist to extract and synthesize them.

Barnacles were passengers long before ships sailed the oceans. Humpback and grey whales bear large white patches of thousands, some of the hitchhikers measuring several inches across. The larvae pick up the scent of their host as they float through the sea and move towards it. Then they dig into the skin—and the great beasts pay a price in energy as they drag their hangers-on through the water. The whale retaliates by replacing its skin three hundred times faster than humans do, in an attempt to slough off its passengers. Some marine mammals secrete enzymes that dissolve the glue and help keep their foes at bay, while grey whales swim into shallow water to try and scrape the unwelcome passengers off. Dolphins move fast enough to wash off such visitors before they can affix themselves. Big sharks, which idle through the water, are nevertheless also free of the pests, because their skin is covered with tiny ridges that prevent barnacles from attaching themselves. An artificial coating that mimics its structure is now being tested on ships' bottoms.

Other species of barnacle hitch lifts on the gills of fish, bore into coral reefs, or live around the deep sea vents that belch out hot, rich water that nourishes a thick living soup, just right for a filter feeder. Not all cirripedes have a settled way of life, for some float blithely through the seas and never touch a solid object.

The most aberrant kinds take up a sinister profession. From a whale's or a matelot's point of view, a barnacle is an irritant but little more. For crabs faced with a marauding cirripede, things are much worse. A certain group lives as parasites within their living bodies. Their macabre habits give an insight into the spectacular diversity of form that evolution can come up with when it generates variations on a body plan.

The barnacle's takeover begins when a female larva lands on its victim and finds a soft spot in its armour. Then she stabs it with a hollow needle and fires a few of her own cells through. She dies at once, but the baleful blob finds its way to the lower part of the crab's body and sends out fine tendrils that run through the host's entire anatomy. They grow to make a mesh that looks more like a mould than a marine animal, and suck in food. The crab seems to stay healthy and continues to eat as fast as it can to feed the visitor. When the time is ripe, the parasite opens up a small hole to the outside and awaits the arrival of a mate—a male larva. The male repeats the actions of its predecessor, inserting its spiny self through a hole and sealing it to prevent the entry of a rival. Now the crab is in real trouble. The male parasite fertilises his partner and she begins to pump out thousands of larvae. The host's whole economy is hijacked and it can no longer grow, shed its skin, or even replace a damaged part. Instead it devotes its energies to its inner barnacle, sometimes for years.

Soon the crab, male or female, is spayed by its unwelcome visitor. A castrated male crab starts to look and behave just like a female. Either sex now acts as a mother—but a mother who cares for another's interests. The host develops a pouch on the underside that resembles that made by a healthy female just before she releases her offspring and spreads them through the water with sweeps of her claws. The unfortunate crab now waves a mass of newborns on their way, but they are not its own progeny but barnacle larvae ready for their next target.

Such diversity confused Victorian biologists, who could see no logic in cirripedes' variety of shape and habit. Much of Darwin's work turned, in the traditions of the time, on an attempt to understand how the various species are related to each other and to find where the group as a whole fits into the animal world. That pastime—taxonomy, as it is called—was once little more than stamp collecting, but it became the raw material for a deep insight into biology. His classification was based on the solid plates that surround most settled barnacles, and he persuaded himself that he could see a hint of order that reflected their family ties (even if he stayed confused by his Chilean burrower and by the parasites). The scheme has been greatly modified, and a system based on the pattern of adult plates remains ambiguous.

The new philately—molecular genetics—uses DNA itself to say much more about shared descent. The assumption that mutations accumulate at a regular rate can generate a family tree of any group's evolution. Well-dated fossils can then, in principle, be used to measure how fast the changes happen, to give a molecular clock. The barnacle tree suggests that stalked forms came first and the others followed. The double helix also hints that the protective armour emerged after a jointed-legged animal had taken the decision to settle down and wait for food to arrive rather than going out to hunt for it. Several of the classical groups, such as the naked barnacles that so confused Darwin, appear to be a mixed bunch with distinct origins, and even the rock barnacles are an assorted lot (although the deep-sea vent types and the parasites are each a group of true relatives). The genes also confirm his view that the cirripedes as a whole fall neatly into the larger family of crabs and lobsters—the crustacea—and into the wider clan of insects, spiders, and other jointed-legged animals. Some biologists claim, on the basis of shared molecules, that insects themselves are a specialised group of crustaceans that reached land. If so, they make an unexpected family that unites barnacles with butterflies.

Whatever the details of their relationships, the diversity of cirripede life began long ago. Two of the great evolutionist's books deal with their fossils. They are not, perhaps, the most riveting of his works but make nevertheless a forceful case that today's kinds descend from forms long extinct. Darwin referred to modern times as the "Age of Barnacles" and at least in terms of the numbers of known species he was right. Cirripedes do not appear in the rocks in any numbers until the demise of the dinosaurs, sixty-five million years ago. A few spots do reveal good evidence of their passage. The Red Crag deposits of East Anglia were laid down in the cool Essex seas of two million years ago. The rust-coloured rocks are still full of the protective plates, mixed in with snail shells and the teeth of the largest sharks that ever lived. Impressive strata in southern Spain record, in the mix of their cirripedes, the rise and fall of a vanished sea. The petrified memorials of whale barnacles show that those creatures, too, have been around for at least two million years, for a bed in Ecuador is filled with their remains as a hint that the whales once bred there. A 164,000-year-old whale barnacle specimen from a settlement in an African cave also shows that humans have long eaten those huge mammals. They scraped off the external parasites and cooked them.

Only a few really ancient examples have been found. A fossil from three hundred million years ago looks rather like a modern barnacle. Another well-preserved remnant found in Herefordshire from a hundred million years earlier—about the time of the first land animals—resembles the larva of a modern cirripede. The low-growing adult forms found on rocks, abundant as they now are, emerged much later, perhaps no more than 140 million years ago, when *Archaeopteryx* walked the Earth.

Anatomy, genes, and fossils all place the barnacles in close association with crabs and lobsters, and in less intimate kinship with insects, spiders, and more. That larger family appears very early in the record, in the Cambrian, more than half a billion years before the present, the era in which life first left abundant evidence of its passing. Some of the mysterious creatures with bizarre body plans found just before that time (and once claimed to represent a unique and vanished fauna) may in fact have been early crustaceans. A molecular clock of the whole group puts the origin of the barnacle lineage well back into the Cambrian (although none of their remains have yet been found) or even earlier. If the clock can be trusted, the first cirripedes may have emerged as part of the vast explosion of diversity that began at just that time among jointed-legged animals and is still evident today.

What sparked off the barnacle big bang? Why did they, like their crab and insect brethren, evolve such diversity of form? And why, much later, did vertebrates, the group to which we and the barnacle goose belong, do much the same? Backboned animals are less diverse in their body form than are cirripedes, but they include creatures as different as mackerel, toads, pythons, and vultures. Why was their evolution, like that of barnacles, so radical, while groups such as sponges or flatworms remained tediously conservative? The answer began to emerge from Darwin's labours over the Down House microscope.

Its owner was the first to identify a barnacle larva, from his strange shell borer from Chile. As he dissected more and more species and examined their juvenile forms, a great truth began to dawn: that the creatures were far more distinct from each other as adults than they were during their early stages. From Scottish rock dweller to naked Chilean, and from tasty marine snack to the sinister enemy of crabs, the young of the various species were remarkably similar. Even better, they looked much like the equivalent phases in crabs and lobsters. Darwin's excitement at this discovery is manifest: he writes of a larva "with six pairs of beautifully constructed natatory legs, a pair of magnificent compound eyes, and extremely complex antennae." His children laughed because the sentence read like a newspaper advertisement by a cirripede manufacturer, but he knew he had hit upon a crucial piece of evidence for evolution.

Most barnacles release thousands of tiny fertilised eggs into the sea.

Each goes through a series of stages, in most cases while floating free in the plankton. The first has jointed limbs attached to a soft, flattened body. The young animal has an eye spot, sensitive even to dim light, that allows it to choose the level at which it floats. Soon it develops jaws and antennae and starts to feed. It goes through several moults and in time turns into a strong-swimming form with a tough outer coat. Those mature larvae prefer to stay near the surface, do not eat, and may be carried far from where they were born. They must find a place to settle down, or—as almost all do—they will die. Some stumble upon a rock, a whale, or a crab and glue themselves on with their antennae. The rock- or whale-dwelling species put out a chemical message—a protein hormone—that invites others to join the colony. For them, every visitor is welcome, for a male must be within penis length of a female if he is to have a chance to pass on his genes, and the more there are, the better his prospects.

Although the early stages of many species look very similar, some—like those that amused the children of Down House—have aberrant juveniles, adapted to their own special way of life. Those of the burrowers cannot swim but scuttle about on the bottom using their antennae as feet. Crab parasites have abandoned the first few stages altogether and hatch as jawed and hungry forms that search for new victims. Natural selection is, it seems, at work on the larvae, which have to adapt themselves to nature's challenges just as grown-ups do. Even so, the young reveal much more about the family's affinities than do the much modified adults. They show how cirripedes and their relatives are a theme with variations.

The same is true of the embryo on a wider stage. That of a barnacle goose is almost indistinguishable from the contents of a vulture egg, and an embryonic human looks much like an embryonic mouse (or indeed, if seen early enough, an early goose). What emerges into the world is quite distinct from what appears early in development. Now we understand why.

Adult cirripedes (apart from the crab parasites) are—like lobsters and insects—arranged in obvious sections, with a head and a thorax divided into six segments (although they lack an abdomen, found in nearly all their relatives). We do not often think of ourselves as segmented creatures, but the vertebrate body is, like that of a barnacle or a lobster, also based on a series of distinct units arranged from front to back. The human head, thorax, and abdomen are obvious enough, but our muscles and our braincase show little sign of order. A glance at the embryo, though, shows that men and women, like their submarine relatives, are constructed from a series of modules, neatly arranged in early life but shuffled around and modified as growth proceeds.

The fossil remains of our watery past as primitive fish, together with the juvenile forms of our living relatives among fish, snakes, and birds, say more. They show how the building blocks have multiplied and rearranged themselves to make today's complicated creatures.

Just three of the thirty or so great divisions of the animal world are organised in obvious segments: the worms, the jointed-legged beings such as insects, spiders, lobsters, and barnacles, and those with backbones. For each, a subdivided way of life has been an evolutionary triumph.

Segmented creatures played a large part in the Cambrian Explosion. Fossils from that time show how the addition of new pieces to a simple body, like beads on a string, can spark a burst of change. Many were wormlike or had jointed legs and external skeletons. Over time they added more sections, evolving into a wild diversity of forms. One ancient marine group, the trilobites (now extinct) started off with around eight segments. Later, some kinds ended up with a hundred, and others with three. That process then reversed itself for some reason, and at the peak of their success most trilobites had no more than thirty-five separate elements.

Barnacles and their relatives have been through the same process of increase, decrease, and divergence. Darwin persuaded himself that the archetypal crustacean, the ancestor of both cirripedes and lobsters, was based on twenty-one parts, divided among head, middle, and rear end. Many modern species have six sections in the head, six in the thorax, and five in the last, abdominal, part. Some have multiplied and modified particular pieces while others have done the opposite. Lobsters, for example, have many more paired and jointed appendages—legs and swimmerets, plus others used in mating and in brooding young—than do crabs, while the barnacles appear to lack the whole rear segment of the body. They are the Manx cats of the crustacean world (and are hence an analogue of the first birds, which were dinosaurs that shook off their tails).

Goethe—philosopher, scientist, and author of *Faust*—had, well before the *Beagle* voyage, noticed hints of pattern within the bodies of fish, birds, and mammals. He came up with a universal theory of anatomy, based on the notion that vertebrae—the individual sections of the backbone—were units from which many of our various parts derived. (The leaf, he thought, had the same role in plants.) Goethe saw life as a biological Proteus based on simple components that could be multiplied and modified into a diversity of structures, the skull included. He was wrong in the details, but his idea has an element of truth.

Although it is simplistic to claim (as Darwin never did) that "ontogeny recapitulates phylogeny," that living beings relive their ancient history as they develop from the egg, the embryo is a reminder of where we came from. The shift from the formless ball of protoplasm that is a fertilised egg to man or woman seems complex but is, in its basics, straightforward. As in origami, a limited set of instructions persuades pattern to emerge from simplicity. As an embryo folds itself into being, its past unfolds before our eyes.

Hints of order emerge early. A fertilised egg divides to form a ball of cells, which in time turns itself inside out and attaches to the wall of the uterus. It lengthens, and a ridge—which soon becomes a tube, the precursor of the spinal cord and brain—forms along the upper surface. The masses of tissue on either side then begin to break up into a series of evenly spaced blocks called somites. Those near the front appear first. Although no physical evidence of their presence can be seen earlier, special stains show that ordered structures arranged from front to back are present well before the somites themselves become visible.

Those units in their rows look simple, but they give rise to complicated structures, some of which have no obvious regular pattern: vertebrae (which would have pleased Goethe), ribs, muscles of the back and the limbs, skin, tendons, and even certain blood vessels. The organised nature of vertebrae is obvious enough, but to the untutored eye the muscles of the leg or the skin on the back give no hint of segmentation. Even so, they—like many other organs—began as blocks.

As development goes on, the front half of one somite fuses with the back of the somite ahead of it to form the precursors of vertebrae—the repeated elements of the spine, a structure common to fish, frogs, snakes, birds, and humans. They surround the spinal cord with a flexible sheath which solidifies as bone is formed. The process is controlled by growth factors, which sometimes go wrong. That has an echo of Goethe, for after a failed attempt in the 1960s to conserve his corpse his body was stripped of flesh, to reveal that the great poet suffered from a debilitating fusion of several spinal bones.

How can a uniform tissue break up into blocks and then into distinct organs? In 1891 William Bateson—the rediscoverer of Mendel's work—came up with a "vibratory theory of the repetition of parts": the notion that a flow of chemicals did the job. Just as waves on the sea create ripples on the sand, their equivalents in the body stamp order onto disorder. A century later, he was proved right.

As the embryo develops, chemical signals promoting growth diffuse from its rear end and move towards the front. They are matched by a second molecular message that travels in the opposite direction and tells the growing tissue to mature and stop dividing. Each potential somite has an internal timer that instructs genes to work for the appropriate time and then to switch off. When the signal arrives, the clock starts. The somites each contain a hundred or more genes that cycle in and out of phase with each other, many with opposed effects on cell division, growth, and movement. Together they build the block of tissue—and the genes that do this job are remarkably similar in mice, chickens, and barnacles, proof that the basic rules of segmentation were in place before they last shared an ancestor.

Vertebrae still retain strong hints of their segmented history. The numbers vary from species to species. Most people have thirty-three (although several are fused together), geese have more (particularly in their necks), and snakes may have over five hundred. The vast increase among the serpents arises because the clock within each of their somites ticks several times faster than does our own. As a result, the undifferentiated tissue is converted into many more segments in the time available—and the animal gains its long and flexible backbone. Perhaps the same is true of the goose's neck.

Each human vertebra has its own personality. Some are reduced to form a vestigial tail, and others fuse to make a solid block at the lower end. Those in the upper part of the back grow great spines to which muscles are attached, while the seven vertebrae in the neck are specialised to move the head. Whatever its task, each has, as a reminder of a shared embryonic experience, a strong resemblance to its neighbours.

The skull, at first glance, looks different. Its twenty-two bones show no obvious segmentation and (apart from the lower jaw) are fused together. The cranium is a round case with many openings and a variety of special structures such as the eye socket, the teeth, the jaws, and the ear. It seems to have little in common with the backbone upon which it perches.

Once again, the embryo is the key. The skull is in fact partly built from somites, with the rest formed from bone laid down by precursors of other tissues. The genes prove that parts of at least the first two somites contribute. As further evidence, mouse mutations that damage the signalling machinery also affect the cranium. The skull, complicated as it is, begins as just another building block in the body's support axis. Its anatomy, its fossils, and its genes show how segmentation can make complex structures from simple precursors.

Every part of the head refutes the "argument from design," the ancient and threadbare claim that complex organs require a designer. Darwin himself used the eye as evidence against that notion, but the ear makes an even better case. Fossils join embryos to show how evolution has cobbled together a series of solutions from what is available. A designer who did the same would lose his job.

The human ear has an outer, middle, and inner section. Together they pick up vibrations from the outside world. The outer ear receives sound waves, the middle amplifies them with a set of bony levers, and the inner ear turns that mechanical energy into pulses of liquid and finally into electrical and chemical impulses that pass to the brain. The inner ear also gives its owner a sense of physical position and of acceleration or deceleration

The organ in its intricacy is witness to the power of variations on a theme. Genes, embryos, and fossils all show that it evolved from the skeletons of early fish—and that it shares some components even with the sense organs of barnacles.

All land vertebrates have some form of ear. The outer ear, or pinna—that elegant appendage on either side of the head—is rather new. Frogs, reptiles, and birds do not bother with it. The structure is made from the same cartilage and skin as much of the rest of the body surface. Darwin noted that in humans and apes, unlike dogs, it could not move, perhaps because as large tree-climbing animals they had less need for eternal vigilance. He was told by "the celebrated sculptor, Mr Woolner" that while working on the figure of Puck, the artist had noted that some people had a small protuberance folded in from the outer margin—perhaps, he thought, a vestige of a formerly pointed ear. The structure is now known as Darwin's point.

The membrane of the eardrum is the gateway to the middle ear. When sound strikes, it vibrates and passes the energy to three tiny bones—the hammer, the anvil, and the stirrup, each named after its shape—that act as levers. Each fits into the next, and together they amplify the movements of the drum into larger movements that are passed down the chain to a small membrane-covered window on the surface of the liquid-filled inner ear. Because the eardrum is much larger than the window, the system increases the pressure upon the gateway to the inner ear by twenty times. As a safety measure, tiny muscles attached to two of the bones damp down the harmful effects of loud noise. The inner ear in turn transforms the physical energy of sound waves into electrical messages about intensity, pitch, and direction that pass to the brain.

The three-part middle ear is as characteristic of the mammals as is hair or milk. All other land-living vertebrates have only one bone in that part of the organ. It is a wonderful example of how repeated structures can be used for a diversity of ends. Fossils, embryos, and living animals combine to paint the picture of how a body built on modules can adapt to a new challenge.

The ability to detect waves in air or water began long ago. Most marine creatures have simple sensory cells that give them that talent. Fish are fairly unsophisticated. Living in water, an excellent conductor of wave energy, they manage with just a series of pressure sensors on either side of the head and body. Land animals need to amplify the feeble power of waves in air. They use the middle ear to do so. Fossils show that the structure appeared around two hundred fifty million years ago, a hundred million years after the direct ancestors of mammals split off from their reptile ancestors, and seventy-five million after the bird and lizard lineages diverged from the same source. All three lineages developed a bony lever independently, and with its help each improved its ability to detect high-pitched sounds. Reptile and bird eardrums connect, rather inefficiently, to the inner ear via just one bone, the stirrup. Mammals' unique set of triple levers, in contrast, much improves their hearing. Each of the bones can be traced to simpler and more ancient structures.

Anatomists uncovered the first hints of the history of the ear at about the time Darwin began his barnacle work. They saw that in the early phases of development, just after the somites appear, the embryos of fish, reptiles, and mammals generate a series of looped arches on either side of the front end. Those six repeated structures grow into matched pouches on each side of the developing head.

Four hundred million years ago, the only animals with backbones were flat-headed fish that swam in an immemorial sea. Those primeval vertebrates had bodies covered in bony plates and ate without benefit of jaws. In time they were succeeded by fishlike creatures with necks and the beginnings of limbs. These in turn clambered ashore around 365 million years ago and in time evolved into frogs, lizards, birds, and people. The fossils of those antecedents of the vertebrates tell the story of the middle ear. It confirms that told by the embryo and by the genes.

In early fish, the arches were supports for the gills, the structures that extract oxygen from water. They did that simple job for millions of years. As their descendants moved onto land natural selection began to modify the arches' repeated structure. In time the arch nearest the front became the first jaw. The lower and the upper jaws of all vertebrates, one hinged into the other, hence trace their origin to an ancient aid to fish respiration. The second arch was picked up to make a bone that connects the upper jaw to the braincase. As their descendants crawled onto land, that structure evolved into a lever to amplify sound.

Lizards and their descendants the birds retained only that single bone. As the immediate precursors of modern mammals appeared, the ear began to commandeer other parts of its ancestors' anatomy. First, the position of the hinge between the upper and lower jaws shifted compared with that found in reptiles. As it did, it freed one bone within the upper jaw and one within the lower. Those newly redundant structures became the hammer and anvil of the middle ear—which means that we hear, in part, with what our ancestors chewed with. Fossils of early mammals as they began to evolve from their reptilian ancestors some three hundred million years ago reveal the whole process, in all its steps, with a series of creatures with more and more complete middle ears. The shift from food processor to hearing aid happened several times in different mammal lineages, most of which are now extinct. Those small creatures of our earliest days ate insects and moved around at night—when any improvement in hearing would have been useful indeed. Anatomy agrees about the evolution of the ear: the nerves that serve the stirrup bone branch from that structure to the face, while those to the other two bones are offshoots of a different nerve (a fact otherwise baffling).

Each of the three bones of the middle ear hence comes via a different route from two of the fish gill arches. Those ancient building blocks have also been taken up for other ends. In mammals, remnants of the first arch help make some of the chewing muscles. The second evolved into some of the muscles of the face and into a bone in the neck that supports the tongue and is important in speech.

As the embryo develops, the famous arches can be seen to reinvent themselves to become parts of the middle ear. The genes that build them, too, resemble others still active in the gill slits of modern fish. The case for the middle ear as a pastiche of an ancient marine structure is watertight.

The inner ear, deep within the skull, is another legacy of an extinct fish—and even of an early barnacle. It too reveals its history in fossils, embryos, and DNA. The sea is a noisy place, for water is almost transparent to sound. Whales sing, fish grunt, and crustaceans join in; the pistol shrimp gains its name from the loud clicking of its claws, while its relative the mantis shrimp (whose claw can break a fisherman's finger) emits a deep rumble that frightens off predators. Lobsters make alarm signals by scraping their antennae across a ridged section of carapace. The larvae of lobsters and crabs—with their close resemblance to those of barnacles—pick up the roar made by waves crashing upon a reef, and make their way towards the sound from kilometres away. Fish are even more responsive.

All three groups use the same fundamental mechanism: not an ear but a set of specialised pressure-sensitive cells on the body surface filled with jelly, into which is affixed a hairlike structure that extends to the outside. A passing wave—a current, the echoes of a surf-battered shore, or the movements of a nearby enemy—causes the hair to flex and the cell to pick up that movement, transform it into electrical activity, and transmit the information to the brain. Our own inner ear has exactly the same arrangement. The physical movements of the middle ear bones make waves that disturb a set of sensitive hairs, which in turn generate a nervous impulse. Damage to a certain gene causes deafness—and a search through fish DNA finds the same gene active in the pressure-sensitive cells. So similar are the two systems that fish are now used to test drugs that might as a side effect damage hearing in humans.

As Wagnerians can attest, human ears do rather more than detect simple changes in pressure. Our ability to tell musical notes apart, impressive as it is, emerges—once more—from expansion and diversification. The simple pressure-sensitive cells of fish have evolved into a series of sensory cells with different sensitivities to particular tones, multiplied and arranged in order within a coiled bony structure. The reptile version is short, allowing them to hear only low sounds, that of birds intermediate, and the mammal inner ear sensor the longest of all. The story of the ear is one of multiplied structures modified for a new and different end. Perfect pitch, for those who have it, has been reached by most imperfect means.

DNA shows that the modular plan identified in the embryo goes back to long before the evolution of barnacles, geese, or men. Whole sections of the double helix were multiplied or lost as evolution made crustacea, birds, and mammals. The first hint of their existence came from certain fruit-fly mutations that double the numbers of wings or antennae and are due to changes in the genes that control the passage from embryo to adult. Such homeobox genes, as they are called after a short repeated sequence (or "box") found in all of them, alter the timing and rate of growth of various segments and change the shape of what they build. They are a molecular mirror of Darwin's discoveries among the barnacles: of duplication, reshuffling, and deletions of parts. As they multiply, the sections diverge to take up new tasks (and on the way destroy another plank of the creationist argument: that evolution can only remove information and not create it).

One surprise in modern genetics was how small the molecular divergence among animals actually is. A goose and a chicken are almost identical at the DNA level, and neither is very distinct from a human. Barnacles are close to the crabs and not very different from flies. Many of their genes have not changed at all in the millions of years since they diverged. Geese and barnacles look different—just as cars and aeroplanes are distinct, although each is built from the same basic elements. Genes make the nuts and bolts of the body. What they make is put together in different combinations for different jobs as developmental switches activate or suppress particular genes in the embryo. The evolution of segmented animals turns, as a result, on the gain or loss of homeobox genes.

In some creatures they are arranged in the same order as the body parts (head first, then the middle section, and then the abdomen), but that neat arrangement is often disrupted as the homeoboxes are broken up into clusters or scrambled altogether. Different creatures have from four homeoboxes to four dozen or so. Their presence in barnacles and buzzards, sea urchins and squirrels, spiders and snails, suggests that the universal ancestor of all those creatures was a segmented wormlike being in an ancient sea, with around eight of the famous genes.

Our own are arranged in four groups with about ten members each. Many are arranged in order—strings of related structures, each specialised to its task. Some help build the ear and are—as the fossils predicted—related to those that in other creatures give rise to gill slits, jaws, and fish sensors. The vertebral column, too, is the product of a set of such genes.

The simplest extant member of the vertebrate clan is a small marine creature called the lancelet that spends much of its time buried in sand in shallow seas, filtering food through its jawless mouth. It has a segmented body and, instead of a proper backbone, a simple rod along its back. The animal's homeoboxes are arranged in the same order as its body parts. Many of its relatives (ourselves included) have four or more times as many such structures. Such multiplication promoted the wild diversity of animals with backbones (and the bony fish, who have eight times the lancelet's allotment, are the most variable of all vertebrates in size, shape, and way of life).

The vast variety of the crustacea and their relatives—from spider crabs to funguslike parasites to wasps—also emerge from this group's flexibility at the homeobox level. Variation upon a common theme reaches a peak among the cirripedes. Compared with their close relatives the lobsters they seem simple, for they have no obvious abdomen and only a few jointed legs. Like snakes, lizards, and whales, the barnacles have lost limbs, and, like birds in relation to dinosaurs, they have abandoned their back ends. The parasitic forms are even simpler. All this can be tracked to changes in the structural control genes. The ancestral barnacle had ten, each of which has an analogue in geese and humans. The numbers of legs vary from species to species—and that variation is matched by the activity of two homeoboxes. The absence of an abdomen is due to a deletion of a group of genes similar to those that code for our own posteriors.

Darwin's "unity of type" hence stretches from cirripedes to men and includes the intimate details of the genome itself. Homeobox genes unite animals that at first sight show almost no resemblance to one another. Sir Robert Moray was, in a way, almost right about the barnacle's relationship to the goose: beneath their superficial resemblance is a shared pattern of repetition—of vertebrae in the goose and of body segments in the goose barnacle. What at first glance might seem an accidental resemblance is proof of an ancient unity of form. Bird and barnacle each show how multiplication and divergence rule the world of life. They put to rest the absurd idea that complexity demands design and that evolution cannot generate information. The anatomy of those creatures, the pressure sensors of fish, and the ear of an opera fan are, like large parts of the human genome, messy and expedient solutions to a set of immediate problems. As Darwin noticed on the coast of Chile, and as modern genetics can affirm, biology is inelegant, redundant, and wasteful. Even so it works well—but only as well as it must.

CHAPTER 8

Where the Bee Sniffs

A GIFT OF ORCHIDS IS A STATEMENT OF A GENTLEMAN'S INTENtions towards a potential partner. A man willing to spend so much on his mate must be devoted indeed—or rich enough not to care, which often comes to more or less the same thing. An orchid, with its extravagant flowers and price tag, is a test of his readiness to invest in a relationship.

The plants feel much the same. Their Latin name, the *Orchidaceae*, means "testicle," after the interesting shape of their roots. Orchids advertise their sexual prowess with expensive and often bizarre blooms. So great are their carnal powers that the English herbalist Culpeper called for caution when they were used as aphrodisiacs. In *The Descent of Man and Selection in Relation to Sex* Charles Darwin showed how, in the animal kingdom, the battle to find a mate was as formidable an agent of selection as the struggle to stay alive. Males, in general, have the potential to have far more offspring than do members of the opposite sex—if, that is, they can fight off their rivals and persuade enough females to play along with their desires. Losers in the battle reach the end of their evolutionary road. Their genes go nowhere. Natural selection as played out in the universe of sex is as pitiless as that which turns on survival. Sexual selection can lead to gigantic antlers, a vivid posterior, or—for species interested in such things—fast cars and Armani suits.

He also examined the sexual struggles of the second great realm of life, the plants. He showed how the search for a partner is even more of a challenge for a flower than it is for a peacock. The sexual habits of the botanical kingdom were obscure (and their existence often denied) until the seventeenth century, but within a hundred years or so the basic machinery was worked out. Flowers were the site of the reproductive organs and a statement of erotic needs. Darwin found that they had arisen in much the same way as an animal's sexual advertisements and were subject to the same evolutionary forces. Their structure was often equally bizarre. For plants, as for animals, sex was full of dishonesty and discord, with everyone involved ready to cheat whenever necessary.

Botanical marriages are more crowded than their animal equivalents, for a third party is needed to consummate them. For some species, wind or water step in to help; but most flowers need a flying penis—a pollinator—to carry their DNA to its goal. Darwin saw how antagonism between plants and their sperm delivery service is as powerful an agent of selection as is the balance of female choice and male competition that gives rise to the peacock's tail (although his acquaintance John Ruskin advised his female readers not to enquire "how far flowers invite or require flies to interfere in their family affairs"). Flower and pollinator, trapped in each other's embrace, enter an evolutionary race that ends in structures as unexpected as anything in the animal world.

The interests of those who manufacture the crucial DNA and those who carry it are different. From the point of view of a female flower (or the female part of a hermaphrodite plant), one or a few visits by a winged phallus is enough to do the job (although the more callers she gets, the greater choice she has of which sex cell to use). To beat its rivals, though, a male is forced to attract the distribution service again and again—and that can be expensive.

In his 1862 volume *On the Contrivances by Which British and Foreign Orchids Are Fertilised by Insects and on the Good Effects of Intercrossing*, Darwin studied the divergence of interest of flower and pollinator. He used the showiest and most diverse of all plants as an exemplar. As he wrote, "The contrivances by which Orchids are fertilised, are as varied and almost as perfect as any of the most beautiful adaptations in the animal kingdom." As well as an exhaustive account of the structure and relationships of the orchids themselves ("I fear, however, that the necessary details will be too minute and complex for any one who has not a strong taste for Natural History"), his work introduced the idea—much developed in *The Descent of Man and Selection in Relation to Sex*—that parts of evolution turn on the ancient and endless sexual conflict that crafts the future of all those drawn in.

The war between flowers and insects is an overture to a wider world of biological discord. It leads to spectacular bonds between improbable partners and reveals many details of the mechanism of evolution (including its uncanny ability to subvert the tactics of an opponent). The orchids and their pollinators were an introduction to the dishonesty that pervades the world of life.

Pollination has attracted attention since ancient times. Both Aristotle and Virgil were interested in bees, but only because they made honey (or collected the stuff, for the Greeks thought it fell from the skies: "air-born honey, gift of heaven") rather than because they were essential for reproduction. The Egyptians, in contrast, understood that dates would not grow on cultivated palms unless male flowers were shaken onto the females. They used their slaves as pollinators. Many other creatures have been called in as marital aids, and their need to do so drives their own evolution. Two hundred thousand insects are known to transfer pollen, and their explosion of diversity began soon after the origin of flowers. From the tropics to the sub-Arctic hundreds of species of birds are also busy shifting genes. Some, such as humming birds, can never afford to stop, as they need a constant supply of nectar to keep their tiny bodies active. Mammals are involved, and one Ecuadorian bat has a tongue half again as long as its body—in relative terms the longest of all mammalian tongues. It stays coiled up in a special cavity in its chest, except when its owner is feeding. One African tree is even adapted for pollination by giraffes. In Australia, marsupials have taken up the job. There, the honey possum has lost many of its teeth and gained a long tongue.

Plants want their go-betweens to be cheap, trusty, and eager, while pollinators would prefer to be fat, wanton, and as idle as possible. The flower shows that a reward is on offer while the other party must decide whether the work needed to get it is worth the effort. A bunch of flowers is an advertisement—a silent scream from the sexually frustrated. Like all advertisements it attempts to reassure those who see it that a high quality product is on view. However, in life as in commerce, the temptation to cheat is never far away; to make false promises with no reward, or to take the prize and fail to complete the task.

Plants and animals make signals of many kinds. They advertise their qualities as mates, their willingness to fight for space or food, and their ability to escape a predator (who might as a result be dissuaded from bothering to attack). Why are the signals so often honest when the reward for dishonesty is so high?

Some are impossible to fake: large tigers make scratch marks higher up a tree trunk than their smaller rivals and as a result hold bigger territories.

Often, though, the information is indirect: a black and yellow wasp warns predators that it is dangerous without needing to sting them all. Such secondary signals can be expensive. Giant antlers, vivid tails, or spectacular blooms are made only by those who can afford them—the healthiest, the sexiest, or the most aggressive. As a result, much of what we interpret as the joys of nature costs a lot: stags die in battle, and a male nightingale loses a tenth of its body weight after a night spent serenading a potential partner. Testosterone itself, that item of masculine identity, is pricey. It suppresses the immune system. A red deer stag in sexual frenzy is open to attack by parasites—and if he can keep roaring in spite of his tapeworms he might have particularly fine genes. Elephants go further. Now and again, one falls into a state of "musth," in which his testosterone level goes up by fifty times and the agitated beast becomes so aggressive that a small elephant will fight to the death against a larger rival.

Flowers, too, are far from cheap. Orchid fanciers pay tens of thousands of dollars for prize specimens, and the trade as a whole has a worldwide turnover of several billion. The orchids themselves invest more of their limited capital into sexual display than does the most enthusiastic gardener or sex-obsessed elephant, for if they do not their evolutionary future is over. The cost of sex to each orchid (and to those who market them) is manifest in the fact that, in the world of the garden centre, many specimens are grown from cells in culture rather than by persuading the plants to go through the expensive rituals of reproduction.

Orchids are animated billboards. Their signals, like those of stags, wasps, or elephants, involve two parties: those who transmit the message and those who receive it. Each needs to know whether the information is accurate and whether the plant with the brightest blooms or the wasp with the most obvious warning is in fact the most generous or noxious. The system is constantly tested by fraudsters. Often they do well: black and yellow is no more difficult for an insect to manufacture than is brown or blue—and a whole group of harmless flies does just that, with bright stripes that falsely advertise a waspish sting.

Swindlers also flourish in the botanical world, among orchids most of all. Darwin found that dishonesty was rife among those elegant flowers. Many specimens had gorgeous displays but gave no payment to their pollinators. He found it hard to believe that Nature could be so fraudulent or insects so foolish as to fall for "so gigantic an imposture" and suggested, wrongly, that his plants provided an as yet undiscovered reward. His observation throws light on a question that he failed to solve: how can natural selection favour the false? The orchids give the answer.

The battle for sex is a war of all against all. It often ends in an arms race, a conflict in which every move made by one party is countered by the other. Sometimes, as in the Cold War, each antagonist is forced into massive investment, and, as in those days, negotiation can end in stalemate. To an untutored eye that may look like peace, but it is no more than battle deferred. The orchids, beautiful as they are and exquisite as their adaptations to the needs of their pollinators might be, show such a struggle hard at work. They prove that propaganda—false information—is useful in both love and war.

Many of Darwin's observations were made on "Orchis Bank," close to Down House, where he found eleven species of the plants. He also studied specimens sent to him from all over the world. He soon saw how the conflict between plant and pollinator would lead to change for both parties. He writes of an orchid: "the *Angraecum sesquipedale*, of which the large six-rayed flowers, like stars formed of snow-white wax, have excited the admiration of travellers in Madagascar." It had "a whip-like green nectary . . . eleven and a half inches long, with only the lower inch and a half filled with very sweet nectar. What can be the use, it may be asked, of a nectary of such disproportional length? . . . in Madagascar there must be moths with probosces capable of extension to a length of between ten and eleven inches! . . . As certain moths of Madagascar became larger through natural selection in relation to their general conditions of life . . . those individual plants of the Angraecum which had the longest nectaries . . . and which, consequently, compelled the moths to insert their probosces up to the very base, would be fertilised. These plants would yield most seed and the seedlings would generally inherit longer nectaries; and so it would be in successive generations of the plant and moth. Thus it would appear that there has been a race in gaining length between the nectary of the Angraecum and the proboscis." In 1903 that long-tongued insect, product of an endless contest with its plant, was at last discovered and named Morgan's sphinx moth. A long conflict of interests had forced both flower and pollinator to adapt to the other's demands.

As the sage of Down House collected orchids from local fields and heaths and examined those sent from afar, he became more and more impressed by the ingenuity with which they pass on pollen: "Hardly any fact has struck me so much as the endless diversities of structure,—the prodigality of resources,—for gaining the very same end, namely, the fertilisation of one flower by the pollen from another plant." He glimpsed but a small part of the game played by all plants as they fulfill their sexual destiny.

As Darwin discussed nine years after the orchid book, in his book on sexual selection in animals, a male peacock's flashy rear says nothing about the merits of tails but a lot about his status as a high quality mate, able to afford a gorgeous adornment. The same is true of plants. More food allows them to make more blooms and proclaim their excellence to a larger audience. The brightest and most generous individuals get more pollinators and pass on more of their genes, promoting yet more brightness and generosity in the next generation and, almost as an incidental, leading to an outburst of diversity as the balance of sexual advantage shifts in different ways in different lineages.

Orchids are members of the great subdivision of flowering plants that generates just a single leaf as the seed germinates. Their kin include the grasses, bananas, tulips, and more. The orchids themselves are among the largest families of all and only the daisies and sunflowers are provided with more species. Around twenty-five thousand different kinds are known—about one-twelfth of all kinds of flowering plant—and no doubt more remain to be discovered.

Because they are so attractive, orchids are important in the conservation movement (cynics call them "botanical pandas"). Although they look fragile, many are tough. Their capital lies in the wet and cool hill forests of the tropics, and a third of all known species are found in Papua New Guinea. Plenty more live in the Arctic and in temperate woodlands, fields, and marshes. They grow on the ground, high in the branches of trees, or on rocky slopes and grasslands. A few live underground and never see daylight. Some prefer dry places and, like cacti, develop thickened stems or tubers to store food and water. Certain kinds have leaves as big as those of bamboos, while others are parasites with almost no foliage at all. One or two, such as the vanillas, make vines sixty feet long. A few are tiny, with a flower head that would fit on the head of a pin. Plenty of others have multiple displays several yards long, and the flower of one tree dweller from New Guinea has a girth of forty feet and weighs almost a ton (and a specimen of that plant caused amazement at the Great Exhibition in 1851). The occasional orchid has opted out of the endless and expensive conflict with animals and is pollinated by the wind, while one Chinese kind has abandoned the whole business of sex and indulges in a strange internal dance in which its male element curves backwards and inserts itself into its own female orifice.

Some of the flowers are simple. They are dark and look rather like the entrance to a burrow, which attracts a bee to come in for a snooze and pollinate as it does so. Others use more elaborate tactics. A few are perfect six-pointed stars, while others resemble a glassblower's nightmare with fine tendrils hanging in delicate and lurid bunches. Yet others look as if they were moulded from thick, pink plastic. The flowers are scarlet, white, purple, orange, red, or even blue. One species is pollinated by a wasp. It generates a chemical identical to that emitted by a leaf chewed by grubs—the wasp's favourite food. The visiting wasp gets not a meal of tasty flesh but a load of unwanted pollen. For those overimpressed by the beauties of botany, certain orchids smell like putrid fish.

The biological war between flower and insect involves—as evolution always does—endless tactics but no strategy. It has produced an astonishing variety of blooms, each of which emerged in a manner that depends on the preferences of pollinators and on what turns up in the way of mutations. Darwin noted how well the orchids provided evidence against the then prevailing notion that the beauties of nature emerged from some kind of design: their structures "transcend in an incomparable degree the contrivances and adaptations the most fertile imagination of the most imaginative man could suggest."

However remarkable their details, all orchid flowers have the same fundamental plan. It resembles that of the distantly related but simpler lily. The parts are arranged in threes, or in multiples of that number. The central lobe is sometimes enlarged into a coloured lip, which acts both as a flag to attract insects and as a landing strip that allows them to reach the sweet reward at its base. Often, the flower rotates to turn upside down as it develops. The male organ sits at the end of a long column, and the male cells, the pollen, are not powdery as in other plants but are held together in large masses, with up to two million minute grains in each. The female part lies deeper within on the same column. Once fertilised, most orchids produce thousands of tiny seeds in every capsule—only a minute proportion of which have any hope of germinating.

When the pollinator enters, some means is found to attach male sex cells to it. Many orchids have a spring-loaded mechanism that fires a mass of pollen in the right direction. It sticks on with powerful glue. As Darwin found by stimulating the flowers with a pencil, the stalk of the transferred pollen sac soon dries out and the clump of male cells takes on a vertical position, ready to fertilise the female part of the next plant visited by an insect. In one kind, should the mass miss its target, its energy is enough to carry it a metre from the plant (it is "shot like an arrow which is not barbed"). The blow is sufficient to cause an insect that has been hit to avoid, if it can, the male (but not the female) part of the subsequent flowers it comes across (which improves the sexual prospects of the male that scared it off). In other species, the pollen masses crack open only to the noise of a buzz like that of a particular species of bee. Yet other kinds have a seesaw that tips the insect onto the pollen mass. However the job is done it pays the plant to do it properly, since for many orchids a shortage of visitors limits their ability to reproduce.

Once the pollinator has been enticed to turn up, it expects to be paid. The first reward of all, in the earliest flowers, was pollen itself, which is expensive as it contains lots of protein. Plenty of orchids still provide a solid meal of the stuff, or of bits of tissue that resemble it. Others give nectar, which is simpler and can be provided in very dilute form. Honey bees, for example, must extract nectar from several million flowers (a few of which may be orchids) to make a single pound of honey.

Some of the rewards are intimate indeed. Certain bees are so tied to their botanical partners that their own sex lives have come to depend upon them. They obtain their sexual scents—their pheromones—from an orchid flower, and without a visit cannot mate. The pheromones may have more than a dozen ingredients. As in the Chanel factory, the bees practice "enfleurage": they mix an odoriferous base taken from the plant with an oily substance of their own that helps the smell to persist. A special grease is smeared onto the flower and the sexual mix is transmitted to pockets in the bees' hind legs. The arrangement evolved from the insects' ancient habit of marking their sexual readiness—like dogs around lampposts—with scents extracted from flowers, rotten wood, and even faeces.

More than half of all orchids are tied to a single pollinator. Fossil water lilies from ninety million years ago have flowers quite like those of their modern ancestors—evidence that their association with beetles dates back at least that far. However, to become too closely connected may be risky. Darwin speculated that the giant orchid of Madagascar would disappear if its specialised insect died out, and he was probably right.

Fidelity does not always last. The battle between animal and plant is not evenly matched, for the insects—many of whom visit a variety of flowers—are under less pressure to retain an accurate match than are the orchids. Orchids evolved long after insect pollination began and have often had to adapt to the needs of their partners. Some insects are quite catholic in their tastes, and some orchids are pollinated by more than a hundred different kinds. Particular flowers do tend to concentrate on similar insects: long-tongued bee flies and long-tongued flies, or tiny bees, flies, and beetles, which pick up the pollen on their legs. Even the bees that take their sexual scents from an orchid are less dependent than they seem. One South American species has become naturalised in Florida—where its normal partner does not grow. It finds its chemicals instead in aromatic plants such as basil and allspice when it chews their leaves and extracts the smelly substances. It pollinates a wide variety of local plants, which reciprocate with nectar if not with an aphrodisiac. Once again, the insect has more freedom of action than does the plant. The orchid's ability to impose itself on its ally is further limited because such gorgeous beings are often rare and scattered among other species. Make life too hard and the pollinator will sip elsewhere.

As a result, the two are less entangled than Darwin imagined. Molecular trees of relatedness of plants and pollinators suggest that the insects have often switched from deep and demanding flowers to species with shallower flowers from which nectar can be sucked with less effort. Infidelity by the pollinator is bad news for the orchid, as it may fail to export its own genes and in addition may get pollen from the wrong species.

The pressure for sex has often caused natural selection to run away with itself. Like many showy animals (birds and butterflies included) orchids have evolved lots of varieties. Twenty thousand kinds are known, versus no more than a hundred or so wild roses—which are happy to attract almost any insect that passes by. Most of the barriers to gene exchange among orchids are held in the minds of their pollinators. As a result, gardeners have been able to generate hybrids by circumventing the bond between flower and insect with a simple paintbrush. Their success shows how finely balanced is the barrier to the movement of DNA. In some flowers, a mutation that changes colour from a hue attractive to bees to one favoured by birds has started a new species in a single step. In orchids pollinated by scent-seeking bees, a subtle shift in the proportions of each constituent can attract different kinds of bee, which means that physically identical plants may in fact be distinct entities that never exchange genes.

Orchids bolster the Darwinian case that new kinds of plant and animal arise through the action of natural selection. He realised that their diversity had been driven by the vagaries of insect behaviour but remained puzzled by the origin of flowers themselves, which he called "an abominable mystery" and a "perplexing phenomenon." The mystery has been cleared up and the orchids have helped.

Plants colonised the land more than four hundred million years ago. Those pioneers had no flowers, and neither did the giant ferns that covered large parts of the planet a hundred million years later. The fern forests declined and the dinosaurs flourished in a world without blooms. Not until the first flowers emerged, perhaps a hundred and fifty million years ago, did the conflict between insects, orchids, and the rest begin. The spectacular joint transformation that resulted led to more than three hundred thousand species of flowering plants and several times that number of insects.

The oldest known fossil flower, from a famous bed close to the estuary of the Yellow River in China, dates from around 125 million years ago, a time when the white cliffs of Dover were being formed in a shallow sea. It looked rather like a water lily and floated in fresh water, blooming above the surface. For tens of millions of years such structures remained modest, but sixty-five million years ago—just as the dinosaurs left the stage—the planet blossomed.

The orchids began about then and played their part in beautifying an unpeopled world. A distinctive pollen sac has been found attached to a stingless bee in twenty-million-year-old amber from the Dominican Republic. That plant's modern relatives use the same group of insects to transfer their male cells. A tree based on DNA suggests that the fossil appeared quite late in the group's history, and the molecular clock, like that fossil, also suggests that orchids as a whole originated when the dinosaurs disappeared. Their massive radiation happened just after that memorable event and was accompanied by parallel change in the insects that pollinate them.

The great blooming was the first skirmish in the war between orchid and insect. Conflict between plants and pollinators is expensive and never more so than when it escalates. In war, as in love or commerce, lavish display is a test of merit. A military parade intimidates the enemy, and a pricey advertising campaign is a sign of a well-endowed company. The medium becomes the message, the powerful stay in charge, cheats go bankrupt, and, for much of the time, truthful ostentation prevails. The best signals are too expensive to copy, which is why McDonald's sues anyone who imitates its golden arches and why Japanese Yakuza participate in a ritual in which they cut off their own fingers.

The interaction between plants and pollinators is also a matter of economics—and economists may help scientists to understand it. Signalling theory tries to explain how decisions are made when the information available is not perfect—what used car to buy, who to hire for a job, what flower to visit. One test is to look for a reliable and expensive sign of quality, whatever it might be. An applicant for a job in a bank might have an advanced degree in genetics. Useless as that certificate may be to a prospective financier, it is at least a frank—and expensive—statement of overall merit. The system works well as long as everyone stays honest. Sometimes they do not. Simple fraud—a forged Harvard degree—can usually be picked up, but what of a parchment from one of the many so-called universities that advertise their wares worldwide? For example, for a fee of $149 an operation that calls itself the University of Dublin, California, claims that it will mail you an undergraduate diploma based solely on your "life experience"; there are no academic requirements, as the organization offers no curriculum. How can employers differentiate Redding University (another American "life experience" degree mill) from the University of Reading (a respectable institution a few miles west of London)? Thousands of people have such qualifications, and if too many degrees turn out to be false, the whole system breaks down. The risk is real: nine-tenths of the jewellery advertised as Tiffany for sale on eBay is fake, and the company has spent millions trying to get rid of the sellers, who cause huge damage to its brand. If the bogus continue to prosper, the entire jewellery market may collapse.

A study of the economic implications of such false signals (or "asymmetric information," as financial experts call it) won its authors a Nobel Prize. Plants and animals have done such sums for years. Most of the time they get it right and honesty more or less prevails. Sometimes, though, the cheats get in; for if the reward is great enough and the penalty for swindling not too stringent, natural selection can favour fraud. Plenty of pollinators are duplicitous. Insects gnaw into a flower while humming birds poke a hole in its side to gain a prize at minimal cost. Even legitimate pollinators like honey bees turn robbers at once when someone else has broken in. For them, dishonesty pays whenever a chance turns up. The flowers themselves face the temptation to invest in display rather than product—in showy blooms that offer no reward. All this means that the price of sex is eternal vigilance.

Orchids have a whole range of misleading lures. Any pollinator that falls for them gets nothing in return. Some fool their visitors with blossoms that resemble female insects such as bees or spiders. The flowers are larger than real females and may emit a hundred times more of their sexual scent.

Males—understandably—try to copulate with them, and in their failed attempt to pass on their DNA they do the job for the plant. The amatory experience is futile but intense, as many of the befuddled males produce copious amounts of sperm (which costs them a lot to make). Darwin found it hard to believe that a bee could be so stupid as to frot a flower, but in the world of sex, stupidity can pay. A naïve male bee, with females in short supply (as they often are because males emerge earlier in the season), is well advised to travel hopefully because he might arrive; he should copulate with anything that looks even a little like a member of the opposite sex on the off chance that, now and again, he will be lucky. The bees oblige, and the orchids reap the benefit.

Other orchids exploit pollinators' aggressive, rather than their amatory, instincts. They mimic a male insect rather than a female—which prompts the annoyed local territory holder to try to drive out the supposed intruder. Yet more take advantage of the visitors' greed. They advertise not sex but a free meal, but provide nothing. That baffled Darwin. It was "utterly incredible" that "bees . . . should persevere in visiting flower after flower . . . in the hope of obtaining nectar which is never present." He suggested instead that the empty flowers had hidden reserves, which the insects would reach if they made a hole and sucked the plant's juices.

Life is less honest than he imagined, and the flowers really were cheats. About a third of all orchids act in this underhand way—flashy signal but no food. Other plants do the same (often with a few "cheater flowers" on an otherwise honest individual) but the orchids are the real confidence tricksters, making up nine-tenths of all flowering plants known to fool their visitors. DNA shows that the habit has arisen again and again within the group—but it does not always pay. Some orchids that now provide a generous recompense to their pollinators evolved from species that once led a dishonest life.

False flowers—like the harmless flies that look like wasps—often turn to mimicry, with a more or less accurate resemblance to other species that do make a reward. They flaunt a badge of quality to attract an assistant on the cheap. Sometimes, their copies are uncannily faithful. Certain Australian orchids, for example, look like mushrooms and are pollinated by fungus gnats. A few even make small orange and black spots on their flowers, which attract aphid-feeding flies. More often, however, the displays are merely general statements of reward that attract a variety of visitors. The parasite joins a guild of locals whose various species share a resemblance and attract the same mix of insects. Honest plants pay the price when pollinators avoid them after an anticlimactic experience with a cheat. Some orchids—like some mimetic butterflies—are doubly duplicitous: individuals vary in colour one from the next, allowing them to parasitize a wider range of victims. Experienced insects become cynical. They move away more quickly—and fly further—from empty flowers than from those with nectar.

Cheats tend to grow scattered among their hosts, for a group of fraudsters living close together is soon detected by the pollinators. They do best at fooling visitors that have just emerged into the wicked world and have not yet learned to steer clear of double-crossers. As a result such orchids tend to flower in the spring rather than later. In other cases a shortage of pollinators foolish enough to revisit a dishonest plant forces it to make a long-lasting flower that (unlike that of most of its fellows) survives for weeks or months. One Australian orchid uses the opposite strategy: all the plants open on the same day of the year, giving pollinators no time to learn about the gigantic fraud about to be perpetrated upon them.

Although orchids are the real experts, plenty of other associations between plants and pollinators have been subverted by natural selection. Wild peas and beans often make nutrient-rich rewards that attract birds to spread their seeds, but some of their offerings contain nothing of value even though they look like a tasty meal. Yuccas—those spectacular flower spikes of the American desert—are pollinated by a certain moth, which carries a bundle of pollen to the female, inserts it in the right place, and then lays its eggs within the flower. When they hatch, the larvae feed on the maturing seeds and, once adult, they fly off to pollinate another yucca. Close relatives of such moths, though, eat the seeds without bringing pollen.

Fraudulent orchids and pollinators are an introduction to the wider world of sexual dishonesty. When it comes to the need to pass on DNA, animals are as devious as plants (although few can match the orchids). Plenty of animals are bullies that boast of powers they do not possess, or braggarts that claim sexual prowess but in truth are feeble. An ability to roar even when filled with parasites, or a willingness to die in the battle for a mate, is hard to fake, but as in the orchids, apparently dependable advertisements of quality can be subverted.

Many males bring a gift of food to persuade a female to copulate with them. Dance flies, hairy-legged predators of wet places, form swarms during the mating season. In some species, each male brings a gift of a dead larva and mates with his female while her attention is diverted by the meal. Once the bribe has been eaten, the male is pushed off. Other species prolong the sexual experience by wrapping the gift in a silk purse, which the female must open before eating. At once, a chance for trickery presents itself. Some damselflies make elegant and complex purses that take a long time to open but—like a dishonest orchid—contain nothing. Fireflies are just as devious. Males bring a gift, a sticky mass of nutritious gel that goes with the sperm and is soaked up by their mates. Those who can afford more of the stuff make a longer flash and attract more females. A successful male soon runs out of energy. Some cheat, with a long flash and no reward—but they take a risk, for a certain predatory firefly uses the burst of light to find its prey. A false flasher risks death every time he exposes himself.

Darwin's mystification about the behaviour of orchids opened the door to a world of evolutionary discord. The conflict extends beyond plants and pollinators to predators and prey, pathogen and host, and men and their domestic animals, all of which are locked in an endless and often joyless battle. Such ancient disputes explain why the Irish suffered a potato famine, why some diseases are virulent and others are not, and why the Argentinian Lake Duck has a corkscrew-shaped penis several times longer than its body.

Sexual dishonesty, in particular, is almost universal. Many species of bird appear to live as faithful pairs, but paternity tests show that the great majority are happy to defraud their partners, and that that half—or even more—of the eggs of a particular female may be the scions of another male (often one more socially dominant than her partner). Mammals are even more dishonest. The joys of paternity testing with DNA reveal that a male mammal's sexual displays are often subverted: an apparently feeble individual can sneak in while the top stag is preoccupied by roaring and insert his own genes without bothering to make a noise at all.

Monogamy is rare. Only one mammal species in about twenty (some humans included) seems to indulge in it. Even classic examples of reproductive honesty turn out to be charlatans. The male prairie vole appears to stick by his mate through thick and thin and helps bring up the young. Their happy marriage is based on a certain hormone. On his wedding night a surge of the stuff kicks in and appears to tie the male to his partner for life. One politician saw the vole as proof that sex before marriage disrupts brain chemistry and leads to divorce. The hormone, he said, is "God's superglue." It bonds partners together and could do the same for society. The gene that picks up the hormone in the bloodstream comes in several forms in humans too, and—in Sweden at least—men with two copies of a variant that resembles that of the vole are more likely to be married and have a less difficult relationship than other men.

Unfortunately, the cold eye of the paternity tester has fallen upon the private life of the prairies. DNA shows that beneath the vole's upright social habits lies a dark universe of sexual mischief. One in five offspring of each pair is fathered by a male other than the marital partner, and around a quarter of all males and females have sex outside the household. Voles are socially faithful but sexually fickle; happy to cheat but quick to forgive.

Darwin was surprised by the reproductive fraud found among orchids—but he refused to accept that the same could be true for mammals, humans least of all. In his view of sexual selection, males might be promiscuous and crafty, but females were monogamous: they chose, and males competed for their attentions. Part of that Puritan view was due, perhaps, to the social climate of his day and to his reluctance to shock the female members of his household. In the modern world, dishonesty in sexual relations is less alarming, as many liaisons consist of longer or shorter periods of serial monogamy accepted by both parties—a shift that shows the flexibility of human behaviour and the difficulty of drawing meaningful lessons about our private lives from those of other mammals, let alone of flowers.

There has, nevertheless, been plenty of sexual chicanery in human history. Casanova (himself of uncertain paternity) posed as a soldier, doctor, diplomat, nobleman, and sorcerer to gain the favours of an admitted 120 women (plus, more than likely, many more). He was a great lover and an excellent liar, although (according to a contemporary) he "would be a good-looking man if he were not ugly": his wit, rather than his looks, helped him charm his way into the bedroom.

Nowadays the prospects for hopeful Lotharios have been improved by technology. No longer does a male need to display his talents directly to raise interest in a potential mate. Instead he can say what he wishes about his looks, education, and wealth on an online dating site with no fear of detection, at least until his first meeting with a prospective mate. Tens of millions of people use such sexual aids, and millions of liaisons have resulted from digital romance. Even so, nine out of every ten users—women more than men—are convinced that the world of electronic eroticism is filled with dirty and decrepit Casanovas presenting themselves as young lovers in their prime in the hope of cheap reproductive success.

Their suspicions are misplaced. Surveys of online daters show impressive levels of accuracy in people's self-descriptions, with nearly all saying something close to the truth about age, body build, wealth, education, politics, marital history, and more (although, admittedly, men told slightly more lies about their income and women about their weight). The advertisers disapproved strongly of anyone who did not live up to their claims on a first meeting and swore that they would go no further with them. Deception is not an effective sexual strategy. For men and women, honesty pays and the fraudulent are rejected as partners as soon as they are detected.

In the dating game, on the other hand, there is almost nothing that a bunch of orchids will not at least start to put right.

CHAPTER 9

The Worms Crawl In

THE FIELDS OF BRITAIN ARE CRISSCROSSED BY EARNEST MEN WITH metal detectors. Despised by archaeologists for the damage they cause, the "discoverists," as they call themselves, have turned up thousands of coins, swords, belt buckles, and the like. During Darwin's bicentennial year one Staffordshire enthusiast unearthed three million pounds' worth of gold objects. They were no doubt buried by their owner in a time of danger, but most such relics simply sank from sight. Why?

Charles Darwin, as usual, got it right. The past had been entombed by worms. He sings their praises in his last book, *The Formation of Vegetable Mould, Through the Action of Worms, with Observations on Their Habits:* "The plough is one of the most ancient and most valuable of man's inventions; but long before he existed the land was in fact regularly ploughed, and still continues to be thus ploughed by earth-worms. It may be doubted whether there are many other animals which have played so important a part in the history of the world." His literary swan song discusses the anatomy and habits of those creatures, their intellectual life (such as it is), and, most of all, their ability to disturb the surface of the Earth: to aerate, turn over, and improve the soil, and to bury anything that lies upon it. Although he claimed that he had produced no more than "a curious little book" on a matter that "may appear an insignificant one," the ravages of the plough since its invention thousands of years ago and the damage done to the Earth's surface by today's agriculture mean that the work of the worms is crucial not only to the history of the world but to its future.

The power of such small beings over the fate of objects far larger than themselves shows, once again, the huge consequences that can emerge from what might seem the trivial efforts of Nature. Darwin was aware of the potential of the worm as proof of the might of slow change; as he said of their efforts, "the maxim *de minimis non curat lex* [the law is not concerned with trifles] does not apply to science." They were the final test of his obsession with the cumulative potential of the small, and he was proud of his results. He dismissed the arguments of a Mr. Fish, who denied the animals' talents, as "an instance of that inability to sum up the effects of a continually recurrent cause, which has often retarded the progress of science, as formerly in the case of geology, and more recently in that of the principle of evolution."

The savant's attraction to earthworms started long before he thought of science. In his *Autobiography* he notes that, as a child, he had been so upset by their contortions when impaled on fish hooks that, as soon as he heard that it was possible to euthanize them with salt and water, he never again "spitted a living worm, though at the expense, probably, of some loss of success!" His later studies introduced a new world beneath our feet, gave life to the idea of animals as a geological force, and showed how even simple beings have a rich mental life. His work became the foundation of a science that has now, almost too late, noticed the dire state of much of the world's vegetable mould and begun to do something about it.

In 1837, just a year after the *Beagle* voyage, Darwin presented a paper on earthworms to the Geological Society. Later he published a few notes on the subject, which occupied him at odd moments for forty years. At last, at the age of seventy-two, he wrote *Vegetable Mould*, which was published at nine shillings in 1881, just six months before his death. The book was received with what he called "almost laughable enthusiasm," selling almost as many copies in its first few years as had *The Origin*.

Soil is where geology and biology overlap. Adam's name comes from *adama*—the Hebrew word for soil—and Eve from *hava*—living—an early statement of the tie between our existence and that of the ground we stand on (*Homo* and *humus* also share a root). The epidermis of the Earth is no more than one part in twenty million of its diameter (our own skin, in contrast, is about one five-thousandth as thick as the average human body). Leonardo da Vinci wrote that "we know more about the movements of the celestial bodies than of the soil underfoot," and until *Vegetable Mould* that was still almost true. Since then, earthworms and their relatives have been studied by geologists, ecologists, molecular biologists, and many others. Archaeologists also have reason to be grateful for their efforts, for without them our insight into history would be far less complete, as most of the evidence left by our ancestors would not be buried but washed away. More important, perhaps, without worms we would starve.

Vegetable Mould built upon an observation Darwin had made as a young man. Twelve months after his return to his native island from his famous voyage, he visited his uncle and future father-in-law, Josiah Wedgwood, at Maer Hall in Staffordshire. Wedgwood took him to a field upon which had been scattered, fifteen years earlier, a mass of lime, cinders, and burnt marble, the detritus of his Etruria pottery works nearby. The material had, over that period, been covered by a layer of earth. Wedgwood suggested to his nephew that perhaps worms had done the job. The young scientist agreed, but saw this at first as little more than a "trivial gardening matter." In time, as the notion that small means could give rise to large ends grew in his mind, he saw in those humble creatures a real chance to experiment on the measured workings of nature.

Charles Darwin continued to study the animals as he travelled across England with his family. They were not the only tourists. The Victorians were fond of excursions, and like today's discoverists many were anxious to cart off relics for their own delight. In 1877, during a brief respite from ill health, Charles took his wife to visit Stonehenge. He dug pits around several of the "Druidical stones," as the monoliths were then called, and noted that even the largest had sunk several inches as a result of the work of the worms. Emma worried that her husband might have sunstroke as he sported with the remains, and she recorded her conversation with the site's guardian, "an agreeable old soldier." "Sometimes," the venerable trooper told her, "visitors came who were troublesome, and once a man came with a sledge-hammer who was very troublesome to manage."

A few years earlier, though, a hammer and chisel had been provided at Stonehenge specifically for the use of those who wanted a curio of ancient times. Their intellectual descendants still agitate the nation's soil. A hobby that began with chunks chipped off monuments has become an electronically powered craze, with, at its peak, almost two hundred thousand enthusiasts in Britain (they include Bill Wyman of the Rolling Stones, who markets his own metal detectors for "Treasure Island UK"). For a time during the 1960s, the prime minister, Harold Wilson, was an honorary patron of the discoverists' organisation. The numbers are well down from those frenzied days, and in much of Europe the practice remains illegal, but in 1997 the British government bowed to reality and changed the law to reward those who report their finds (they are, said the culture minister at the time, "the unsung heroes of the UK's heritage"). The Portable Antiquities Scheme, as it is called, applies only to England and Wales, for Scots must still give up their treasures to the Crown. South of the border, the number of objects reported has risen from fewer than a hundred per year to thousands. The Scheme now lists more than three hundred thousand items, the Staffordshire gold the most spectacular.

Some among them were concealed by men anxious to confound their enemies or placate the gods, but many other objects have been hidden by humbler creatures with simpler motives. Electronic sweeps of the fields around Down House reveal plenty of coins, necklaces, buckles, and the like. Vast numbers more no doubt remain to be uncovered. They were interred by worms as the animals searched for shelter and food.

The earthworm has undoubted charm. It belongs to a group known as the annelids, which include the leeches and lugworms, and is related to less agreeable creatures such as the parasites that cause elephantiasis in tropical Africa. It is a more distant kinsman of snails and slugs. Such ancient roots are best revealed by patterns of shared DNA, as soft-bodied creatures leave few fossils (although the remains and the tracks of primitive annelids are found as far back as the early Cambrian). Today some three thousand species are known, and given our ignorance of tropical nature, far more must exist. Most are small and unassertive, but a certain Australian kind grows three metres long and ejects a jet of fluid half a metre into the air when annoyed. Britain has just over a couple of dozen species while France has six times as many. A rain forest has far more.

A 2005 survey at Down House, in the woods by the Sandwalk—the site of its owner's daily stroll—and in a nearby meadow where he had noted that stones were soon buried by worms, revealed that his modest plot was still a hotbed for the creatures. Nineteen species were found nearby. The most abundant nowadays (and no doubt in Darwin's time) was the black-headed worm, which is smaller than the familiar lobworm found in city gardens and used as bait by fishermen. Black-headed worms were particularly abundant in the kitchen garden, probably because of its many decades of compost and a strict ban on pesticides by the present guardians of the house.

A worm is an animated intestine. The body is divided into segments, each possessed of an outer layer of muscle that encircles it and an inner muscle sheet that runs parallel to the axis. Each segment bears a simple kidney. A series of even simpler hearts is distributed along the animal's length. The body is hollow and filled with fluid, with a long digestive tube down the centre. Many kinds have internal glands filled with lime—calcium carbonate. Some have coloured blood while others are almost transparent. Certain varieties smell of garlic, perhaps to put off predators.

The skin is covered with stiff spines that help the creature move through the soil as it eats its way onwards, pumping out waste from the rear end as it goes. In some species, the slime made as the animal burrows hardens into a solid wall that keeps a track open for a possible return, but in others the soil collapses behind the questing worm as it travels. Certain worms live on or just below the surface and in leaf litter while others hide as much as six metres down. A few prefer rotting wood. Those most important to farmers roam the top metre or so of soil. Some kinds reuse their burrows while others set out instead to build new homes. The common lobworm makes a single excavation, with one or two branches, while others make a network with several exits.

Most earthworms spend most of their time at rest in their underground fortresses and venture forth only when conditions are suitable. In winter they dig down and hibernate, and in dry summers they build a cocoon in which to rest until the rains come. After a downpour, the animals may travel in vast concourses across the surface. Darwin noted that many species dislike leaving the doors to their burrows open, and seal them by pulling in leaves. Others make piles of digested earth—casts—on the ground; those of some tropical forms may be several inches high. He also discovered that the creatures excrete with some care, the tail being used almost like a trowel to make a neat heap of ordure. A careful look at the body waste revealed many fine grains of silt that had been broken down from larger particles within the soil.

Worms live no longer than two years or so, and most die much younger. They get us all in the end, but some creatures are able to retaliate first. Badgers and hedgehogs are fond of a diet of worms, and as Alfred Russel Wallace noticed, the native of South America appreciated them too. In the Orinoco Basin of Venezuela, smoked earthworms still form part of the local cuisine.

Many species regenerate their tails when cut off, and a few do the same for a head; but—in spite of myths to the contrary—none of the familiar kinds can develop into two individuals when cut into pieces. (An amputated tail may grow a mirror image of itself, but then it starves.) A few do reproduce by simple fission: the back breaks off and forms a new worm, and—in some—the animal splits into several pieces, each of which gives rise to a new individual. The ability to multiply by breaking into fragments is common in the lower reaches of the animal kingdom, but the worms are the most advanced creatures to possess that talent.

The sex life of annelids is varied. Several are all-female and lay eggs without benefit of males. Some of those clonal kinds spread fast and have invaded new habitats such as sewer pipes. Other species are hermaphrodites, with separate male and female genitalia. Sex happens in a long slime tube in which boy-girl meets girl-boy. The two animals lie head to tail to consummate their relationship. The male checks the virginity or otherwise of its partner's female organs and adjusts the amount of sperm accordingly. It triples the volume when it senses that its mate has already had sex with another, no doubt with the aim of flooding out the previous donation. The animals usually copulate underground, but sometimes move to the surface (in Darwin's words, "their sexual passion is strong enough to overcome for a time their dread of light"). A swollen midsection of the body forms a protective cocoon as the eggs are laid.

Worms are among the simplest creatures to have a central nervous system, with a distinct brain connected to a set of nerve cords (although after the brain has been removed the animals can still mate, feed, and find their way through a maze). One of *Vegetable Mould*'s sections is headed "Mental Qualities," but its first sentence reads: "There is little to be said on this head." Even so, its author set out to see just what his subjects' lowly wits were capable of. He noted that they often pulled in leaves to seal the mouth of the burrow, perhaps, he thought, to protect themselves from the cold. Whatever crosses the animal's mind as it drags fronds into its home, it acts with a degree of foresight. Worms, he found, prefer to grasp a leaf by its tip. More than nine-tenths of broad leaves were pulled in from that end, but for narrower kinds, easier to slip into a small opening, the fraction was just two in three. Rhododendron leaves curl up when on the ground, so that some were narrower near the base and others near the tip. The creatures more often than not pulled them in by the narrow end. They were just as smart when it came to pine needles, which could be dragged in base-first only.

Darwin admired such apparent rationality, for it forged a link between the lowest creatures and the most noble: "One alternative alone is left, namely, that worms, although standing low in the scale of organization, possess some degree of intelligence." In the first real experiment on invertebrate psychology, Charles and his son Horace presented the animals with paper triangles cut into various shapes—and once again they acted in the most efficient way, seizing the pointed end. Other studies of their intellectual universe involved the choice of foods—meat, onions, starch, or lettuce (with beads and paper used in an attempt to trick them). In a series of midnight expeditions to the lawns of Down House, father and son shone lamps upon the animals, warmed and cooled them, and subjected the unfortunate creatures to tobacco smoke. His subjects were "indifferent to shouts" and equally unconcerned by the shrill notes of a metal whistle or the deep tones of a bassoon. But they did respond to vibration, and became agitated when placed on top of a piano. They were "more easily excited at certain times than others," and a series of taps upon the ground made them emerge. Hungry birds could often be seen doing just that to persuade their prey to venture forth. There was, no doubt, a wide gap between their mental world and that of the naturalist—but profound as it was, it had been bridged by the same system of slow change that had shaped their physical universe.

Those patient experiments on the inner life of the burrowers were an introduction to their wider role in the world of the soil and their ability to modify their own habitat and that of those who stride its surface. Most of his book is devoted to the animals' impressive ability to disturb and to fertilise the ground.

That ability had been noticed long before. Aristotle described worms as the "Earth's entrails." Cleopatra decreed them to be sacred animals and established a cadre of priests devoted to their well-being. Her interest arose because the creatures seemed so important to the fertility of the mud laid down by the Nile (they were also useful in weather forecasting). Herodotus knew as much when he wrote that "Egypt is the gift of the Nile," and much of the river's immense deposit that comes down in the annual flood does indeed begin as eroded worm casts in the Ethiopian highlands far upstream. The same is true on Darwin's island. In 1777 the English naturalist Gilbert White wrote (in a letter unknown to his great successor) of their "throwing up infinite numbers of lumps of earth called worm-casts which, being their excrement, is a manure for grain and grass . . . the earth without worms would soon become cold, hard-bound and void of fermentation, and consequently sterile." As he put it in *The Natural History of Selborne*, "Earth-worms, though in appearance a small and despicable link in the Chain of Nature, yet, if lost, would make a lamentable chasm."

Without their help, we would ourselves fall into a real abyss. Those simple beings play a role in both economics and history. They improve drainage and break organic matter into fine particles; as his book put it, "All the vegetable mould over the whole country has passed many times through, and will again pass many times through, the intestinal canal of worms." That unromantic product determines the fertility of the soil, which in turn does a lot to dictate the nature of the society that lives upon it.

Nineteenth century gardeners saw the animals as mere pests, relatives of tapeworms and such unpleasant beasts (the word *vermin* has the same Latin root as does *worm*). Books advised how to get rid of the creatures by driving them from their burrows with mallets, poison, or steel rods inserted into the ground and played with a bow. The production of soil was thought to depend not on biology but on chemistry and physics, on the mechanical dissolution of rock and the chemical decay of vegetation. The Comte de Buffon, mentioned in *The Origin* as a pioneer of the notion of natural selection, was, like his British successor, interested in what made the Earth's outer cloak. He noted that many soils contained grains of minerals such as iron, and that cover tends to be thinner on mountain slopes than on valley floors. All this was proof, he thought, of the breakdown of rocks by acids and of the importance of rain, rivers, and gravity in disturbing the surface. The Russians of the same period—obsessed as they were with the vastness of the steppe and its effect on the Slavic psyche—were pioneers in the study of the deep and dark *chernozem*, the "black earth" that fed the masses and nourished the nation's soul. They too emphasised the role of chemical decay in the sacred soil of Mother Russia and dismissed the importance of such mundane creatures as worms.

Physics and chemistry do help build the ground beneath our feet. Chalk and limestone dissolve in weak acids, and even sandstone and granite can be eroded away to make earth. Tiny cracks fill with water, which shatters rock when it freezes. Whatever the temperature, the surface tension that holds water to the walls of minute channels also exerts huge pressure as the liquid warms and cools. Clay itself—little more than tiny particles of ground rock—is a product of such insidious action.

The earliest known fossil soil is three billion years old, almost as ancient as the land itself. It was made with no help from biology. Three hundred and fifty million years before the present day, the first land plants moved onto an almost sterile landscape. Since then, life has fed on soil and soil on life, in a great cycle that enriches both.

The labour of worms did a lot to improve the Down House garden. Its topmost metre or so is filled with channels, most of them thinner than a human hair. Around half of them are filled with water. Below this lies a sheet of material with little air and no worms. A large part of the animals' contribution comes from their ability to open the ground in this way. An acre of rich and cultivated ground is riddled by five million burrows—which, together, add up to the equivalent of a six-inch drainpipe. Half the air beneath the surface enters through burrows, and water flows through disturbed soil ten times faster than in unperforated.

The surface of the Earth, if watched for long enough, is as unruly as the sea. Everywhere, it is on the move. Gravity, water, frost, heat, and the passing seasons all play a part, but life disturbs its calm in many other ways. Living creatures—from bacteria to beetle larvae and from badgers to worms—form and fertilise the ground. The largest reservoir of diversity on the planet lies beneath our feet, with a thousand times more kinds of single-celled organisms per square metre than anywhere else. A typical patch of earth contains more species than the Amazon rain forest. Its vast variety of inhabitants, large and small, burrow through the topmost layer, bring in air, digest its nutrients, excrete into it, and turn over so much material that the skin of our planet is in constant eruption. The "biomantle," the organic layer near the surface, can be metres deep or little more than a thin sheet. Its base is marked by a layer of pebbles that sinks to a depth at which the stones can no longer be disturbed by the agitations of life. As the biomantle churns, the relics of man's labour—from stone age tools in Africa to the pots of the first European settlers in Australia—sink through the topsoil and accumulate, with the stones, at about the level where the tillers abandon their efforts.

Darwin's subterranean subjects have many assistants. A shovel of good earth contains more individuals than there are people on the planet. Many soils have hundreds of thousands of tiny mites and springtails in every square metre. Roots exude sugars and other substances that feed millions of single-celled creatures. They add their remains to the helpful productions of the worms' rear ends. Bacteria and fungi possess powerful enzymes that break down material that even earthworms cannot digest. They feed roots, break down vegetation—and produce antibiotics. Until the 1930s, a diagnosis of tuberculosis was a death sentence; the disease had killed a billion people since Darwin's birth. Then it was found that a soil suspension killed the bacteria responsible. Soon streptomycin was discovered and the disease was, if not eradicated, at least kept at bay. The microbial world beneath our feet is still almost unexplored. It must have much more to offer. Molecular probes that pick up known genes in unknown species suggest that soil contains innumerable members of a distinct group of creatures called archaea, which look like ordinary bacteria but occupy a separate kingdom of life. They were once thought to be eccentric denizens of a few hot springs, but we now know that some places contain a hundred million in each gram of soil. Each burns ammonia and other waste products and helps maintain the Earth's fertility.

As roots grow, they push barriers out of the way and later die to leave channels into which earth may collapse. As they suck in water, the soil settles, and as trees lash back and forth in storms they further disturb the ground. A large tree can shatter solid rock when it falls, and the hole it leaves may take centuries to fill. Small animals do even more. Insects, mites, spiders, and subterranean snails, together with worms, may make up fifteen tons of flesh in a hectare of soil—as much as one and a half elephants (and a single pachyderm needs several times that area to feed itself).

The elephant under the grass is a voracious beast. Earthworms are earth movers, but in the tropics, ants and termites do more, for they carry material up from several metres down. Alfred Russel Wallace was astonished by the richness of the ground in some parts of Brazil: "a layer of clay or loam, varying in thickness from a few feet to one hundred . . . over vast tracts of country, including the steep slopes and summits . . . of a red colour, and is evidently formed of the materials of the adjacent and underlying rocks, but ground up and thoroughly mixed." It had been mixed by ants.

Larger creatures also help to stir up the soil (and elephants themselves often paw away at the surface). Below ground are moles, prairie dogs, marmots, wombats, meerkats, badgers, and other excavators. They are helped by aardvarks, armadillos, and anteaters as they scratch away in search of food. A colony of naked mole rats can build a network of burrows a kilometre long. In the southwestern United States, tens of thousands of symmetrical piles of earth up to two metres high and fifty metres across mystified historians for years. They were thought to be sacred sites of a lost tribe of Indians. The truth about the Mima mounds is more prosaic. They are built by gophers, which, over thousands of years, pushed tons of material uphill to provide a dry refuge in a marshy place.

Even the bottom of the sea is not safe from disturbance: manatees and narwhals dig up food, skates do the same, and shrimps work away at the top few inches of mud. Enthusiasts for the process trace it back to the Cambrian Explosion, around five hundred and forty million years ago, when the first animals with hard shells emerged. They were able to dig into the thick layered mat of microbes that had until then covered the seabed. A whole new way of existence could then spring into being. The revolution of the burrowers marked the origin of modern life, and their descendants play a central part in keeping it healthy.

Today's worms are merchants as well as miners, for they are major players in the vast traffic in chemicals that passes from the world of life to that of death and back again. Darwin wrote that "all the fertile areas of this planet have at least once passed through the bodies of earthworms." In an English apple orchard they eat almost every leaf that falls—two tons in every hectare each year. In the same area of pasture, they can munch through an annual thirty tons of cow dung. A few tropical species pile their casts into mounds twenty centimetres high, and *Vegetable Mould* refers to the gigantic castings on the Nilgiri Hills of southern India as an indication of the vast amount the animals must chew. Most of their endless meal is ground down in the animals' muscular gizzard. Rather little is absorbed. Even so, it undergoes chemical changes. The experimenter fed some of his subjects soil laced with red iron oxide powder and noted that it lost its colour when excreted; acid and enzymes, he thought, had done the job. The creatures' potent guts do change what passes through them, for the chemistry of clay is much modified when passed through their bodies. It is ground finer and finer, which helps it retain water and nutrients, and its tiny particles mean that clay has ten thousand times the surface area of an equivalent volume of sand.

Many species of worm also have the unexpected ability to draw carbon directly from the air and convert it into soluble substances that can be recycled. *Vegetable Mould* suggests that the small grains of chalk found in worms' digestive glands are waste products. The truth is more remarkable. Radioactive labels show that the glands extract carbon from free carbon dioxide—abundant beneath the soil—at a great rate (an unusual talent for an animal) and combine it with salts of calcium. The particles of chalk thus produced are excreted and return to the earth when the creature dies.

The constant flood of slime pumped out as they burrow also recycles other minerals such as nitrogen. Plants and animals die, and farmers pour manure and treated sewage onto their lands, and the worms do their bit to pull them into the earth. Their casts contain five times as much nitrogen and ten times as much potassium as does the soil itself. Much of that is due to their busy inner life—to the bacteria that live in the oxygen-free world of the gut. Each intestine is a tiny fermentation chamber in which bacteria chew up manure. They make useful fertilisers—but with the side effect that they also pump out nitrous oxide, a greenhouse gas (and, as "laughing gas," a primitive anesthetic), which gives their hosts an unexpected role in global warming.

A simple experiment shows the worm's disturbing power. A mouse carcass was placed in a glass jar with some fine rubble and leaves, plus one earthworm. In just three months the bones had been scattered sideways across about ten centimetres and had been dragged the same depth into the soil. In wormless jars the corpse stayed undisturbed. Darwin himself set out to test his subjects' powers of burial. On morning after morning, in the garden at Down House, he counted the number and size of casts—each the undigested remains of a worm's meal—and found dozens in a typical square yard. His cousin Francis Galton joined in and, ever keen to use statistics, counted the dead worms he saw on paths in Hyde Park. He found, on the average, a corpse every two and a half paces. He calculated that the worms brought seven to twenty tons of earth to the surface in every acre of Kent each year. At that rate, worms would lay down a sixth of an inch of topsoil in a twelve-month period. In fact, their labours are even more impressive, for much of what they excrete stays beneath the surface.

The numbers of worms are so huge, and their labours so sustained, that over time they do great things. In a follow-up of his youthful observation at Maer, and soon after moving to his own grand house, Darwin scattered quantities of broken chalk and brick over a nearby field to test how fast it sank. Twenty-nine years later he dug a trench across the site and found most of the chalk buried about seven inches down. The bricks, on thinner ground, took longer, but eventually even they disappeared. By 2005, the fragments of brick had sunk half a metre or so to the level of a solid band of clay into which the worms could not penetrate, while the chalk had dissolved away.

The Down garden had ten or more burrows in every square metre. Given each animal's ability to chew through earth, if all of them acted with equal enthusiasm throughout, the whole mass would be disturbed to a depth of a metre in about five thousand years. That was not always the case, for stone tools of that age are often found at much shallower levels. In addition, many species of worms reuse their burrows, and that economical habit reduces the extent to which they agitate the ground. As a result, an object on the surface may be buried quite fast in its first few decades, but then slow down.

In his final years Darwin started an experiment to test the worms' sepulchral power. He placed a lump of rock—a hefty millstone more than a foot across—in a corner of his lawn. A long brass rod was pushed deep into the soil through a hole in the centre of the stone. The movement of the rock in relation to the rod measured the efforts of the burrowers as they worked away below. In the early days it subsided by about two millimetres a year. Charles died before the experiment was complete, but his son Horace continued the study. Today's stone, admired by the curious, is a copy of the original and has been moved since it was first put in place. Nowadays it sinks more slowly than before. Sir Arthur Keith (who was better known for his role in the Piltdown Man scandal) wrote one of the first biographies of the great theorist of evolution and retired to live close to Down House in the 1930s. He reexamined the sites Darwin had used in his chalk and brick experiments. Eighty years on, the marked stones had descended little more than they had in the lifetime of those who had set them there, as further proof that the worms are most active near the surface.

On Leith Hill, the highest point of the North Downs, Darwin estimated the extent to which the material unearthed by his animals would slide downwards to fill valleys and plains. He found that their castings rolled downhill at such a rate that for a fairly steep slope a hundred yards long, twenty pounds of freshly disturbed earth would be washed to the bottom each year. His estimate is close to those made today and a tribute to the worms' importance as architects of the fertile fields of southern England—and of the hungry pastures on the hills above.

At Down House, the longest-running biological experiment in the world is still under way; but, record-breaking as it is, the worm stone has been in place for only a century and a half. Darwin realised that in the abyss of time, his own observations marked a mere instant. The remnants of ancient structures scattered over England gave him a better chance to test his subjects' powers. In late middle age he began a tour of the stately ruins of England and—ever a busy correspondent—wrote to dozens of people who might give him useful information.

The superintendent of excavations at Wroxeter, near Shrewsbury, came up with a strong hint of what worms could do, given time. The city had been founded by the Romans to act as the capital of a British tribe, the Cornovii. Viroconium, as it was called, at its peak held six thousand people. It fell into decay and became, in legend, one of the several possible sites of King Arthur's court. Camelot, the city's archaeologist wrote, was in some places buried under four feet of vegetable mould. Much of that was due, Darwin had no doubt, to the efforts of earthworms.

In 1877, men at work on the restoration of Abinger Hall, the grand house of his friend Thomas Henry Farrer (who had helped with the experiments on hops and other climbers), discovered the remains of a Roman villa. The Sage of Down came to visit. He saw how the creatures had crawled through the rotten concrete floor of the ancient building and brought up material from below. At the time, and for long afterwards, antiquarians assumed that the layers of earth found above decorated pavements and the like were the remnants of later and less civilised inhabitants who had settled into the houses of their erstwhile masters and left their rubbish behind. The supposed squatters were, in truth, worms.

Darwin was impressed to discover burrows almost six feet beneath the modern surface. The animals could even mine the ancient structure's thick walls. Farrer observed their activities for several weeks and saw them hard at work as they heaved earth to the surface. A quick sum showed that their labours were more than sufficient to bury a Roman house within a few centuries. At a villa with a mosaic floor in the Isle of Wight, Charles's son William was told that so many castings were thrown up between the tiles that the surface had to be swept every day to keep the pattern in view. William also visited Beaulieu Abbey in Hampshire and found that the bottom of a hole dug down to the ancient floor twenty years earlier was already covered by excretions.

The trip to Stonehenge was part of the worm project. It showed that, no matter how active the animals might be in rich soils, in some places they achieved rather less. Emma noted that they "seem to be very idle" and in that thin soil they had managed to sink some of the "Druidical" stones by less than a foot since they had toppled (they rested on the chalk layer beneath, into which the creatures could not penetrate). John Lubbock, who lived close to Down House, had dated the stones to the Bronze Age, which began in Britain around 2100 BC. The latest estimates push its builders further into antiquity at close to 2500 BC—a period when the Britons began to cut down their forests and replace them with fields. Some of the stones fell long ago in part because of the worms themselves, whose efforts and that of the rain have weathered away the soil that once supported them. Others fell—or were pulled down—within the past few centuries (one major collapse happened in 1797), which suggests that perhaps the burrowers were not as idle as Emma imagined. Indeed, they buried the stone chippings left by the first modern excavator of the site in the 1920s to a depth of three inches in thirty years, which was almost the same rate as that measured at Down House.

In Charles Darwin's sesquicentennial year of 1959, a plan was hatched for an improved version of his experimental millstone, built on a grander scale as a test of the destructive effects of earthworms on the monuments of England. The British Association—that great stamping ground of early evolutionists—set up a Committee to Investigate by Experiment the Denudation and Burial of Archaeological Structures. A long pile of chalk with a ditch alongside, of about the size and shape of a typical section of an English burial mound of three thousand years ago, was built at Overton Down, not far from Stonehenge. Plant spores and bits of broken flowerpot were scattered on the surface. Just thirty years later, natural weathering and the efforts of Darwin's favourite excavators had caused large parts of the wall to collapse into the ditch. Both structures were covered with a layer of grass and soil. The pieces of broken pottery moved at a rate of around an inch a decade, and the spores were carried down several inches. Much of this was due to the work of worms. A similar edifice built on an acid heath in Dorset, with far fewer of those animals, was much less disturbed. At Overton, the experimental barrow now looks much like those a hundred times older. Once again, most of the changes took place in the first few years after it had been disturbed. The next survey is planned for 2024, when, no doubt, the British Association Barrow will be almost impossible to tell from those built by the associated British long before.

Life's underground frenzy blurs the record of the past. The diggers at Abinger Hall found several Roman coins; but among them was a halfpenny dated 1715. An incautious student would gain an odd view of British history if he took that observation literally. In a five-thousand-year-old Indian mound in Kentucky, the constant activity of soil animals has been enough to turn over the entire site fifty times since the original inhabitants left. Sites with moist, rich soils are at more risk of disturbance than are deserts or cold uplands—but as men and worms have similar tastes in where they choose to live the news for those who hope to reconstruct ancient history is mixed at best.

As the soil turns over, the casts of those responsible are picked up by water and run away or sink back underground. Wind also moves worm excreta, adding another weapon to the animals' armoury as soil engineers. Windblown soils make up great parts of China, the Great Plains, and the Rhine Valley. A gale moves stones and gravel, but such large elements soon fall to earth, and the finest, most nutrient-rich particles—those from worm casts included—are blown farthest of all. That valuable powder can cross the Atlantic. On the last leg of the *Beagle* voyage the young naturalist noted a fall of white dust onto the deck of the ship as it sailed off South America. Some of that dust came from North Africa. Silt around Lake Chad—in part the product of soil animals—is picked up by gales and sifted finer as it travels, until it consists of valuable salts of nitrogen and phosphorus. More than ten million tons of the stuff fall on the Amazon rain forest each year, bringing fertility to those thin and starving lands. The good work of the worms crosses oceans.

Nowhere is their power better seen than when the animals themselves traverse the seas. Some species—the "peregrines"—are keen migrants. In New Zealand at the end of the nineteenth century, farmers found to their surprise that what had been thin pasture had been transformed into lush loam. The immigrants were at work, breaking down earth into compost. They can move into empty pastures at ten metres per year. In today's New Zealand, as they spread, they can bury metal rings—a modern version of the worm stone—at a rate of ten millimetres a year, five times that in the Down House garden. European worms are now on every continent apart from Antarctica, and in many places they far outnumber the natives.

Their ability to improve the ground is such that they are sometimes introduced to heal the Earth. After mining is finished, or the peat has been stripped from a bog, the intruders can help a damaged landscape to recover. In the Kyzylkum Desert of Kazakhstan and Uzbekistan, vast numbers were moved during Soviet times to isolated oases, with salutary effects. Waterlogged Dutch polders, too, had their drainage improved a hundredfold after the animals were called in to help.

Long before they began their global migrations, worms made the landscapes of the farming regions of the world and maintained gardens in a healthy state. Now, the farmers who till the ground—like their predecessors at the beginning of agriculture—are undoing their work. Like skin stretched thin on an aging face, the earth's epidermis—the soil—has grown thinner with the years. Like age itself, the process is slow but impossible to resist, and like the signs of decay in a human body, the destruction speeds up with the years.

Man has flayed his native planet for ten thousand years. Soil is hard to make but easy to destroy. A modern plough turns over hundreds of tons a day, far beyond the capacity of the most vigorous invertebrate. It digs down no more than a couple of feet, making a solid and impermeable layer at the depth of the blades. When heavy tractors roll across the surface their wheels compact loose earth into something like concrete, in which nothing will grow. Continued ploughing also breaks up the topmost layer and allows vast quantities to wash away. The farmers' raw material is on the move, from hill to plain, from plain to river, and from land to sea. The evidence is everywhere. My parents' house overlooked the Dee Estuary (the Welsh rather than Scottish version). What was, a few centuries ago, a broad waterway has become a green field with a ditch in it, and the local council is much exercised about the rising sand that blows onto its roads. The reason lies in the fertile fields of Cheshire and North Wales. They have been ploughed again and again, and their goodness has disappeared downstream.

The process is speeding up. The amount of organic carbon in Britain's lakes and streams has rocketed in the last twenty years and in some places has doubled. Waters that once ran clear now flow with the colour of whisky (which itself gains some of its hue from the carbon-rich streams that run through peat bogs). On the global scale, things are worse. Twenty-four billion tons of the planet's skin are washed away each year—four tons for every man and woman—and although some is replaced (and some has always been lost to the rain and gravity) the figure is far higher than once it was.

Man has been careless of the deposits in his soil bank. Again and again, as a civilisation grows it empties its underground accounts, goes into decline, and collapses. Usually it takes around a thousand years. Marx noticed as much and wrote that "capitalistic agriculture is a progress in the art, not only of robbing the worker, but of robbing the soil."

Darwin compared the work of the worms to that of the plough. The great naturalist's experiments showed how earth could slip and churn, but he had only primitive tools to measure movement. A sinister spin-off of modern technology has come to the aid of science. Between the explosion of the first atom bomb in New Mexico in 1945 and the last air burst of a hydrogen bomb in 1968, vast quantities of fallout spilled across the world. One constituent, radioactive cesium (with a half-life of around thirty years) binds to soil. In an undisturbed site, the element is most abundant near the surface; but after a plough has passed, the radioactivity is dispersed to the depth of its blade. As the disturbed ground is washed away, the label is lost, to accumulate in places where the flowing mud settles. In the Quantock Hills of Somerset soil is, in this era of industrial agriculture, lost at a rate of around a millimetre every year. As the land sinks, the machines dig deeper and the Roman relics beneath will be smashed within a century—even though, thanks to the worms, they have been preserved for the past two thousand years. Already, most of the treasures picked up by metal detectors come from ploughed fields, proof of how man is stripping the fragile surface.

The Romans themselves paid a price for abusing the ground beneath their feet. At the time of the Western Empire's fall, so much damage had been done to Italy's farmlands that much of the capital's food had to be imported. The local fields lost their fertility, and vast quantities of grain were shipped in from Libya (which itself soon became a wasteland as its surface was denuded). In the Empire's declining years, it took ten times as much Italian ground to feed a single citizen as it had in Julius Caesar's day.

The real damage began long before. The first towns appeared in the Middle East around eight thousand years ago. Quite soon, their growing populations began to demand more food. The farmers exploited their precious soil without replacing its nutrients. They attacked it with ceaseless vigour. The plough was invented soon after oxen were domesticated. In a few centuries the topsoil was gone and so were many villages. Within a couple of millennia, all the fertile land of Mesopotamia was under cultivation. The ground was soon washed away, and irrigation canals became blocked with mud. Enslaved peoples such as Israelites were forced to clear it. The abandoned city of Babylon is still surrounded by great dikes of earth thirty feet high, the remnants of their labours. Abraham's birthplace, Ur of the Chaldees, once a port, is now a hundred and fifty miles from the sea: the plain upon which it sits is no more than the remnants of what were fine fields, lost to leave a desert. Salt soon poisoned the land, and what had been the Fertile Crescent collapsed, along with the civilisation it fed.

Much the same happened in China, where the Great River was renamed the Yellow River two thousand years ago, as swirling earth changed the colour of the water. The ancient Greeks, too, faced the problem when the country around Athens was stripped bare. Plato blamed the farmers.

The real disaster for the skin of our planet came in the Americas. The Maya bled their landscape dry, as did the inhabitants of Chaco Canyon and Mesa Verde—the abandoned Native American settlements, now part of the deserts of the American southwest, whose collapse came when drought was added to the damage caused by cultivation. The Europeans were worse. Virginia's "lusty soyle" was ideal for tobacco—but that plant sucks goodness from the ground as much as it does from the bodies of those who consume it. In just a few years the land was drained of its worth, but farmers saw no need for fertiliser ("They take but little Care to recruit the old Fields with Dung") because they could simply move on to the next piece. George Washington complained that exhaustion of the soil would drive the Americans west (as it did). Charles Lyell, Darwin's geological mentor, used the gullies that scarred the devastated ground of the South to examine the rocks below and commented that soon the nation's farming would collapse.

The real nemesis of soil—and the real threat to our own future—was born in 1838, when John Deere invented the polished steel plough—"The Plow That Broke the Plains," as the memorial at his birthplace calls it. Soon, thousands of his devices were tearing up the prairies.

A thin layer of sod had kept the West's soil in place, and as it was broken to bring more and more of the Plains under cultivation, the ground began to blow or drain away. Now it has lost half its organic matter, much of it down the Mississippi, Mark Twain's "great sewer," which has twice as much silt as it held in John Deere's day. A drought in the 1930s, following decades of shortsighted farming, created the ecological and economic disaster known as the dust bowl. A great gale in May 1934 blew a third of a billion tons of dust eastward from Montana, Wyoming, and the Dakotas. The choking cloud reached New York two days later and finally petered out many miles into the Atlantic. The finest material was blown farthest—which meant that the New Englanders gained nutrient-filled dust that had passed through the guts of northern plains worms, leaving the unfortunate westerners with a rough and hungry silt. The gales returned again and again, until by the mid-1930s more than two million acres of prairie had turned to desert.

Worldwide, an area greater than the United States and Canada has been despoiled by human action. In Haiti, almost all the forest has gone, and much of the nation's ground is now bare rock. Less than a quarter of Haiti's rice, its staple food, is homegrown, and food production per head has gone down by a third. In China, the Great Leap Forward exhorted the peasants to "Destroy Forests, Open Wastelands!" They did, and the soil paid the price. Africa's shared experience explains some of that continent's chronic instability. Three-quarters of the continent's usable land has been bled of its nutrients, and farmers cannot afford to replace them with fertiliser. Their fields are, as a result, no more than a third as productive as those elsewhere. Africa's earth still leaches its fertility into water or dust. The Sahel, the region of thin soil to the south of the Sahara, is losing as much as an inch of surface each year. Hundreds of millions of people go hungry as a result. A planet ploughed by man is far less sustainable than one tilled by Nature.

In 1937, President Franklin D. Roosevelt said in a letter to state governors that "the nation that destroys its soil destroys itself." His Universal Soil Conservation Law became the first step to putting right the damage done to the fabric of his native land. It promoted careful management, the use of windbreaks, and a ban on the reckless destruction of forests. By the 1950s most of the American dust bowl had returned to a semblance of health. To plough with the contours rather than against them makes a great difference.

In China, the Three Norths project will plant a three-thousand-mile strip of trees in an attempt to stop the earth from being taken by the wind, and even in the Sahel there is hope in technology, with lines of stones set across the slopes to stop the thin earth from washing away. In Niger alone, twenty thousand square miles of land have already been put back into cultivation.

Soil protection has a long history. In the Amazon basin, rain leaches the goodness away, and the vegetation feeds on itself, recycling nutrition from its own dead logs. As a result, cleared forests are unproductive. The terra preta lands are islands of deep earth scattered across its thin landscape. They appeared soon after the birth of Christ and were made by native peoples who settled down, fed slow fires with rubbish and leaves, piled up excrement, and added bones to the mix. The New World farmers lived in large but dispersed cities that recycled their waste and set up precursors of the greenbelts that surround many European towns. Over the centuries the ground was filled with worms and their helpful friends. They chewed up the ashes and excreted it as a muddy paste of carbon mixed with mucus. So fertile is terra preta, with ten times the average amount of nitrogen and phosphorus of other local soils, that it is sold to gardeners.

The worm has turned. In many places the plough is out of fashion and the burrowers are allowed to work undisturbed. Farmers scatter seed on undisturbed ground or insert them through the previous year's stubble. The world has a hundred million hectares of "no-till" agriculture, and even in places where the machines still rule, the ground is treated better than before. Brazil, with its ant-built landscapes, is a pioneer. Weeds are suppressed by the mat of dead vegetation that appears, water runs away more slowly, and the temperature of the surface is less variable (and better for seeds) than that of a field made up of ridges of bare earth. Farmers who once saw worms as a pest now realise that to let the animals flourish undisturbed preserves the soil bank far better than their machines and chemicals.

The move away from tractors and towards Nature's tillers is spreading fast. In some places it reduces the rate of loss fifty times over. In Canada, two-thirds of the crops now grow on earth left undisturbed or ploughed in such a way as to reduce the damage. An era in which millions of tons of mud are lost from fields may be succeeded by an age in which plants and animals are once again allowed to regenerate their own environment. Darwin, with his passion for the natural world, would be pleased. Even so, the risk of a global Mesopotamia—a collapse in food production as the worms' precious products are squandered—has not gone away.

In his old age, the great naturalist often spoke of joining his favourite animals in "the sweetest place on earth," the graveyard at Downe. He was denied the chance to offer himself to their mercies and was instead interred in Westminster Abbey, whose foundations kept them out. His remains may have been saved from annelids but have no doubt been consumed by other creatures. Darwin's single and simple idea—of the importance of gradual change in forming the Earth and everything upon it—has replaced the power of belief with that of science. Perhaps today's science of the soil will return the compliment by allowing worms and their fellows to restore the damage done to their product: the vegetable mould that keeps us all alive.

ENVOI

Darwin's Island

WHEN CHARLES DARWIN MOVED TO DOWN HOUSE IN 1842 THE population of England was fifteen million and London, the largest city in the world, had two million inhabitants. At the time of his death, forty years later, the number of Londoners had doubled and the capital's fringes were creeping towards his retreat. Its population has doubled again since then and Darwin's house has become a museum. It sits in a carefully preserved enclave that pretends to be a village but is in fact part of the London Borough of Bromley. A few segments of nearby countryside have been preserved (the "tangled bank," the microcosm of life referred to at the end of *The Origin*, is almost unaltered), but the landscape around Down House is now suburban at best. A glance across the famous Sand Walk reveals the tail fins of planes parked a few hundred yards away at Biggin Hill Airport, a Battle of Britain air base whose pilots claimed to have shot down more than a thousand enemy aircraft. Now it is the most popular light aviation centre in England, and its annual air fair attracts a hundred thousand visitors. A lot has changed in the Kentish countryside since Charles Darwin walked across what has become an oil-stained strip of tarmac.

Many other places associated with the great man—and many of his subjects, from apes to earthworms and from insectivorous plants to *Homo sapiens* himself—have also been transformed. That would be a surprise to the patriarch of Down House. Darwin looked at the past to understand the present. He scarcely considered what the future might bring, for in his view evolution was so slow, and flesh so stable, that no real changes in the world of nature would be visible for many generations. In the long term, no doubt, the outlook was bleak; as he wrote to a friend: "Even personal annihilation sinks in my mind into insignificance compared with the idea or rather I presume certainty of the sun some day cooling and we all freezing. To think of the progress of millions of years, with every continent swarming with good and enlightened men, all ending in this, and with probably no fresh start until this our planetary system has again been converted into red-hot gas. *Sic transit gloria mundi*, with a vengeance."

A moment of hindsight two centuries on shows that he was too conservative in his predictions. Every continent is indeed swarming with men, a few of them enlightened, but progress (if such it is) has not been slow but meteoric. A lot has happened since 1809 and a lot more is about to take place. In the next two centuries, plants, animals, and people will see an evolutionary upheaval greater than anything experienced for thousands of years. Now, the prospect of biological annihilation is far closer than is the certainty of the heat death of the universe.

In the last few weeks of his voyage, in July 1836, the young explorer had a brief vision of what lay ahead. The *Beagle* dropped anchor at Saint Helena, halfway between Africa and South America. The island, first occupied by the Portuguese in the sixteenth century, is one of the most isolated places in the world. Darwin was delighted by it: fifty square miles of volcanic mountain rose "like a huge black castle from the ocean." He admired the "English, or rather Welsh, character of the scenery" and noted with surprise that Saint Helena's vegetation, too, had a British air, with gorse, blackberries, willows, and other imports, supplemented by a variety of Australian species. Over seven hundred plants had been described—but nine out of ten were invaders. They had driven the original inhabitants to extinction or to refuges high in the mountains. A sweep of thin pasture near the coast was known to locals as the "Great Wood"—which is what it had been until the previous century, when the trees were felled and herds of goats and hogs consumed their seedlings and killed the forest. Plagues of rats and cats had come and gone as they ate themselves to extinction. On his first day ashore, he found the dead shells of nine species of "land-shells of a very peculiar form" (one of the few mentions of snails in his entire oeuvre) and—in an early hint of evolution—noted that specimens of one kind "differ as a marked variety" from others of the same species a few miles away. All but one of those molluscs had been wiped out and replaced by the common brown snail of English gardens.

Almost two centuries after his visit, life on Saint Helena is worse. The island then had forty-nine unique species of flowering plant and thirteen of fern. Seven have been driven to extinction since the arrival of the Portuguese, two survive in cultivation, and many more are on the edge. The last Saint Helena olive died of a fungal disease in 1994. Of the tree-fern forest of the high mountains—still in robust health at the time of the *Beagle*—a few parts remain, but other unique habitats Darwin visited, such as the dry gumwood, have gone. Of the ebony thickets just two bushes are left. The island's giant earwig (at six inches the world's largest), its giant ground beetle, and the Saint Helena dragonfly, all common in the 1830s, have not been seen for years, and the snail of peculiar form is now reduced to a population of no more than a few hundred. The Saint Helena petrel is extinct, and just one endemic feathered creature, the wire bird, is left—but that too is under threat.

Three months after the farewell to Saint Helena, the young explorer wrote in his diary that "we made the shores of England; and at Falmouth I left the *Beagle,* having lived on board the good little vessel nearly five years." His account of the expedition ends with a spirited enjoinder to all naturalists "to take all chances and to start on travels by land if possible, if otherwise on a long voyage." Charles Darwin never left British shores again.

He had no need to, for, as this book has shown, the plain landscapes of his own country gave him the raw material for a life filled with science. Darwin's fifty years of work on his native land's worms, hops, dogs, and barnacles changed biology forever. Since then the British Isles have provided another useful lesson for students of Nature, for that modest archipelago is a microcosm of the global upheavals that have taken place since the great naturalist came home.

Bartholomew's Gazetteer, in its edition just after Darwin's death, describes Kent, "the Garden of England," as a paradise: "The soil is varied and highly cultivated. . . . All classes of cereals and root produce are abundant, as is also fruit of choice quality and more hops are grown in Kent than in all the rest of England. The woods are extensive. . . . Fishing is extensively prosecuted . . . of which the oyster beds are especially famous."

A lot has changed. Kentish farms bring in half what they did even a decade ago. The oysters are almost gone, and the salmon fishery of the Thames, which fed the apprentices of London with such abundance that they refused to eat fish more than once a week, has collapsed. Bucolic pursuits have been replaced by that invaluable product, "services," which accounts for three-quarters of the county's contribution to the nation's wealth. Kent is a dormitory of London, and London has become a staging post for the world.

The flow of people, power, and cash has carved up the county's landscape with motorways, rail links, and webs of power lines. The oasthouses that once stored hops have become commuter homes, and the hops themselves—the raw material of groundbreaking experiments on plant movement—cover a fraction of the fields Darwin knew. Although the Great Wen has been kept partly at bay by the Green Belt, plans for a "Thames Gateway" mean that yet more of the Garden of England will soon be a bland suburb.

Much of his work on insect-eating plants, self-fertilisation, and orchids took place in Ashdown Forest in the adjacent county of Sussex, where his cousin Sarah Wedgwood had a house and where he often walked, mused, and botanised. A visit today shows how fast the wild can retreat. In Victorian times the Forest was one of several vast belts of English heath, successors of ancient tracts of trees felled thousands of years ago (Cobbett called it "the most villainously ugly spot I ever saw"). Ashdown was used by the Normans as a game preserve, closed off with a twenty-mile bank. In time most of the trees were cleared and burned (in iron foundries as much as hearths), and it turned into heathland, a seminatural part of the seminatural landscape that is England. Since *The Origin*, the nation has lost nine-tenths of its heaths. The Forest, at a thousand acres, is the largest piece left, but even that is a shadow of what it was. The acid grasslands and marshes have been taken over by bracken or have dried out as water has been pumped away to slake the thirst of millions. Many once abundant species—gentians, asphodels, sundews, orchids, and more—are now rare, and some of the great man's walks have become suburbs, farms, or golf courses.

The Common Plants Survey keeps count of sixty-five of Britain's most abundant flowers, from primroses to bluebells and foxgloves, in five hundred random plots scattered across the country. A century ago those species were almost everywhere. By 2007, a quarter of the sites had none of them. Most of the empty plots were in huge fields of corn or on wide pastures without hedgerows. Others were in woodland. England's forests—albeit preserved by the Woodland Trust, the Royal Society for the Protection of Birds, and the National Trust—have lost much of their diversity. A stable ecology maintained by the labours of woodmen has been replaced by museums of elderly trees in which bluebells and foxgloves, and sparrows, cuckoos, and jackdaws, are in decline. Across England—and across Europe and North America—the fields are starved of life. Subsidies have made a desert and called it farming.

Kent's sorry tale is repeated in many of Darwin's favourite places, from Stonehenge to Shrewsbury and from Wales to the Highlands. In a further blow to the products of evolution, the world has come to Kent, and its animals—and people—have migrated to the world. His home has been united with the rest of the planet, which has become a single giant continent—a global island—rather than an archipelago, real or metaphorical. Humankind, too, has been homogenised, and even genteel Bromley now has a tenth of its citizens from an ethnic minority. No longer does evolution mould the natives of each corner of Earth to fit their own domain. The struggle to exist for both man and beasts has become a worldwide conflict rather than a series of local skirmishes. Some creatures thrive in the international arena, but many more are doomed.

Evolution generates difference. One species and one alone has put the process into reverse. Man has instituted a simplification almost as great as that brought by the explosion that destroyed the dinosaurs. The Galapagos themselves are a stark reminder of what he has done. *HMS Beagle* visited the island of James in 1835. Food was plentiful: "We lived entirely on tortoise meat . . . the young tortoises make excellent soup." In those ungainly creatures, Darwin saw, without realising it, his first hint of evolution, for the animals from that island were distinct from those on nearby Indefatigable and Albemarle. In a rare conjunction of taxonomy with gastronomy he noted that the James specimens were "rounder, blacker and had a better taste when cooked"—which at the time seemed little more than a curiosity but was in fact an introduction to the biology of change.

Now, the tortoises of James and its fellows are almost extinct. From a quarter of a million in the *Beagle*'s day, their numbers have dropped to fifteen thousand. Three of the fourteen unique races have gone, and just one animal, the famous Lonesome George, is left from another (at the age of ninety or so, he was twice persuaded to mate with a female from a different island, but on both occasions the eggs were sterile). Pigs, as much as men, have killed off the tortoises, for they love to feast on their eggs. Goats have blighted many islands. Less obvious pests have also made their way to the archipelago. The cotton cushiony scale insect invaded twenty years ago and has reached across the whole of the Galapagos, attacking dozens of kinds of native plants.

Pigs and scale insects are dangerous because they have epicurean tastes. They are happy to try anything once; and—like the young explorer with the tortoises—will turn to a novel foodstuff if their usual diet is not available. They can hence snack on the last specimens of an endangered species without eating themselves out of house and home. Such generalised predators, as they are called, are a real threat. On the Galapagos, cats are a plague, pigeons have pushed out their feathered relatives, and alien wasps have done terrible damage to the insects. The islands face an era in which specialists, evolved to fit their own small place in nature, are falling to loutish strangers able to cope more or less anywhere. A tourist on the islands today (and a hundred and fifty thousand arrive each year) has less to admire than did the crew of the *Beagle*. Next century's visitors may find the place more or less indistinguishable from South America. The products of millions of years of isolation have been destroyed by man, the most generalised predator of all.

The Galapagos are the icons of evolution and their problems get plenty of attention. Many other oceanic islets—rare, specialised, and fragile as their natives are—face the same cataclysm or worse, but not many people notice. From Saint Helena to Tahiti and from Hawaii to the Cape Verdes, the alarm has at last been raised. It is too late to save such places, most of which began their decline long before the *Beagle* arrived.

The fate of the giant earwig of Saint Helena and the tortoises of the Galapagos is sad enough, but Darwin's more modest subjects provide an equally trenchant statement of the universal attack on the biosphere. They are both under threat and a threat to others. Those humble creatures—the earthworms and bees, the primroses and orchids, the plants that climb and those that kill their prey—all face an ecological earthquake. The lessons they bring are more alarming than are those from the distant isles of the Pacific, for they show that the crisis has moved beyond the exotic to the familiar, and how what was commonplace is becoming rare.

The sundews of Kent and Sussex are far from safe, and many insectivores sent to Down House from around the globe are in deeper trouble. The wide fields of Venus flytraps and pitcher plants that covered parts of the Carolinas have been destroyed. Agriculture and drainage tear up their habitat, and the gardeners who dig them up do not help. A more subtle threat comes from fire control, because such beings thrive best in places often burned—which in today's carefully managed countryside happens less than it once did.

Darwin's other subjects, the orchids, face the same problems. Their enemies are those of the insectivores: aggressive farmers, fragile habitats, and greedy collectors. A third of the fifty British species are under threat, and several have populations of fewer than a hundred—and one, the lady's slipper orchid, was for a time reduced to a single individual in the Yorkshire wilds (thousands of greenhouse specimens have now been sown in the hope of rescuing it). The Victorians suffered from "Orchidelirium," and paid large sums for rare specimens. Traders destroyed whole beds to ensure that their own stock kept its price. A decade after *The Origin* the botanist Joseph Hooker noted how the area around Rio de Janeiro, visited by the naturalist on the *Beagle* thirty years earlier, had been pillaged of its orchids, which have never come back. Unlike the Dutch tulip fever of the seventeenth century the orchid mania is still upon us, with a global trade worth ten billion dollars a year. Expensive specimens sell for thousands. Much of the business is legitimate, and the plants are cultivated or cloned in huge numbers from cells taken from single individuals. Plenty more is not; and many wild species from Thailand, China, Brazil, Guatemala, and elsewhere are at risk. About one species in ten is threatened, and the continued loss of tropical forests means that many more will disappear before they are known to science. Even those on "Orchis Bank," near Down House, survive only because of the vigilance of local naturalists.

The losers in the post-Victorian battle are being replaced by others that thrive in the new global economy. Many have migrated to new lands, where they cause havoc. Weedy plants may be a nuisance, but weedy animals are worse.

Britain itself faces a revenge of the immigrants: a wave of creatures that have appeared from almost nowhere. A New Zealand flatworm introduced to Belfast in the 1960s has run wild (and an Australian cousin has also begun to move). It wraps itself around earthworms and digests them alive. The pest has spread through Scotland, northern England, and Ireland, and in some places worm populations have collapsed.

Modern society has been built upon aliens, creatures moved from their native lands, be they maize, chickens, or cattle. They evolved in the Middle East, in Asia, or in the New World but have been transported to all parts of the globe. Many have become pests in their new homes, and many more have hitched rides with those who cultivate them. Darwin noticed the invasion of British plants into the United States and asked his American colleague Asa Gray, "Does it not hurt your Yankee pride that we thrash you so confoundedly?" The New World soon got its own back, with grey squirrels that eat woodland birds' eggs and Canadian pond weed that blocks streams. The third millennium is the era of weeds, and the weediest species of all—*Homo sapiens*—is to blame.

Plenty of weeds stay at home; they live in disturbed ground, flourish for a short time, and move on to a new patch when conditions change. They do little damage except to the good temper of gardeners. When they escape, they become the botanical equivalents of pigs: they move in, exploit what is available, and destroy the locals. Many imports from the Old World have thrived in the Americas. One European roadside species, the knapweed, a small thistle with a pink or yellow flower, has covered millions of hectares. It secretes a poison that kills native plants, which—unlike those at home—have not evolved resistance. As a sinister side effect it also poisons horses. The knapweed is now out of control. Others, such as the Brazilian water hyacinth introduced to Caribbean islands, find themselves in a place without their native pollinators, take up self-fertilisation, and become a fecund pest. Some even hybridise with a relative, hijack its genes, and gain renewed virulence as a result. The bright yellow and poisonous Oxford ragwort, common in disturbed ground in England, is a hybrid of two Sicilian species brought to the Oxford Botanic Garden in the seventeenth century, which escaped and mated with each other. The new plant is still spreading.

Some of the most aggressive aliens are the climbers. Even the hop has become a nuisance, with a Japanese variety spreading across the United States. Kudzu, a climbing pea, is also native to Japan. It was transplanted into that nation's ornamental garden at the 1876 Philadelphia Exposition. Gardeners liked the flowers and it was dispersed across the country. At first sight, the immigrant seemed helpful. It lays down roots two metres long and in the South was used to reduce soil loss after forests had been cut down. Railroads gave free kudzu to farmers in the hope that they would cultivate it for fodder that their company's trucks could then transport. That was a mistake. The weed grows so fast that the locals recommend, tongue in cheek, that windows be closed at night to keep it out. In some places it extends by thirty centimetres a day—twenty metres and more a season—and can soon smother a huge tree. Kudzu is out of control over an area that straddles Alabama, Georgia, and Mississippi and has spread as far north as Massachusetts and as far west as Texas. Attempts to subdue it cost half a billion dollars a year.

Other climbers are just as busy. Florida has "air potatoes," yams from West Africa that sprawl over trees and block the light. It also suffers infestations of climbing ferns from Asia. English ivy has shaded out tracts of maple forest around Seattle. In Australia the humble blackberry is a nuisance, as is the mile-a-minute vine, a morning glory introduced from the Old World tropics. Most are harmless at home, but a lifestyle that depends on a burst of growth when a sudden open space appears in the forest is lethal when exported to a place not adapted to their wiles.

Some of the climbers' success emerges from another by-product of human activity. The effects of the carbon crisis on climate are familiar enough—but climbers thrive in the new and enriched atmosphere. Over the past two decades the proportion of the Amazon jungle taken up by lianas has rocketed up, in part because the forest has been partly cleared by loggers, but also because of the increased carbon dioxide, which they can soak up and lay down as wood. As a result they flourish at the expense of trees. Ivy, too, now grows at an exceptional rate as it gains extra carbon from the air.

Plenty more of Down House's experimental subjects have found a new home. Even barnacles are a menace. Some, helped by the spread of shipping and by water (larvae included) dumped from ballast tanks, have begun to gallop around the globe. A barnacle the size of a tennis ball, once restricted to the Pacific coast of South America, has within the past five years made its way to Florida, Georgia, and South Carolina. It infuriates boat owners as it acts as an unwelcome brake on their vessels' smooth bottoms. A species common on the coasts of California and Oregon has, in return, spread in huge numbers to Argentina. It was first recorded in the 1970s and has taken over the shoreline for more than a thousand miles. The animal has driven local sea snails, seaweeds, and more to extinction. It has found its way to Japan and at the present rate will soon reach Chile and wipe out the creature that drew the young Darwin's attention to the joys of the barnacle in the first place.

On land, too, a revolution is under way. The worm has turned. Much of the northernmost part of the Americas is almost devoid of native earthworms. They were wiped out by the last Ice Age, which left just a few remnants in the Pacific Northwest and in a few patches elsewhere. Native Americans may deplore what the white man brought, but they did at least import earthworms, carried in pots or on the mud of immigrants' boots. Now fifty and more exotics have arrived. They can advance several metres a year as they burrow, and even faster when they hitch a ride. Escapes from bait tins take them far into remote forests as anglers search for new lakes.

Unlike most invaders, they have been of some use. The West was won with the help of worms. Before the white man arrived, the northern prairies fed no more than herds of buffalo, but as the aliens spread, their fertility soared and maize and cattle moved in. Millions of acres were turned over, and a dense and surly coat of acidic humus that sat on top of a sterile mineral layer was transformed into a well-mixed light soil just right for farming.

Not all the news was good. Before these immigrants put in their appearance, many northern states and much of Canada were covered by fern-filled forests that sprang from deep mounds of leaf litter, or duff, which mouldered over years rather than being dragged into the ground by annelids. The duff sheltered beetles, salamanders, mice, and more. The worms ate it, leaving a naked and unprotected surface. Most of the natives evolved to deal with undisturbed ground and suffer as a result. Thick undergrowth gives way to horsetails and pitcher plants. The local vegetation also depends on a relationship with the fungi that cluster around their roots—and they too have been lost under the assault. Aspen and birch trees die, forests of sugar maple are parched as the water runs through the newly permeable ground, and prairie herbs disappear and are replaced by their European equivalents. In northern Minnesota, great tracts of hardwood have been destroyed in the past forty years, and the problem is spreading. The once ponderous economy of those ancient forests has speeded up, and vast quantities of carbon and nitrogen have been washed away. The problem is not limited to North America, for exotic worms have also invaded tropical jungles. What they will do, we do not yet know.

Insects, too, followed the farmers and some, like the worms, are a help. Most North American bees came from Europe. They arrived within two years of the Pilgrim Fathers, brought in by sweet-toothed pioneers anxious for honey. At once the immigrants set up wild colonies and thrived. So much were they an indication of European settlement that the Indians called them the "white man's fly." A wave of bees moved up the valley of the Missouri at forty miles a year, and, it was said, as the bee advanced, the buffalo retreated. So impressed were the Mormons of Utah by the animal's hard work that they chose it as their state symbol.

European bees still number in the billions in North America. Like the underground invaders they have driven out local species and like them they play a large part in the agricultural economy. Many crops—fruit trees above all—need pollinators if they are to thrive. In fact, 80 percent of the top hundred or so food plants, responsible for two-thirds of global production, depend on them. California has so many almond, cherry, and apple orchards that natural pollinators are unable to keep up, and a healthy hundred-and-fifty-million-dollar annual honeybee rental industry has grown up. As the season moves on, the hives follow the flowers. Without European bees the yield of fruit, seeds, and nuts in the United States would drop by nine-tenths. The aliens may have damaged their local relatives, but they have done a lot to help the people who brought them.

On both sides of the Atlantic bees are in decline. The numbers of wild colonies of European bees in parts of North America are a tenth, and of those in hives a third, of what they were half a century ago. Insecticides, air pollution, mite parasites, viral disease, and competition from introduced African bees have been blamed for the crisis, as have the loss of hedgerows and other scrubby places which were a home for useful insects. In spite of attempts to keep the pollinators happy with wild flowers planted around orchids and by controlling insecticide use, the decline goes on. In an airborne twist to the tale, pollution kills the scent of many flowers and further reduces the chance that an insect will find its target. That news is bad for honey lovers, worse for farmers, and may be catastrophic for many native plants.

The troubles of the bees, the spread of weeds, and the destruction of the soil are just part of a global crisis of agriculture. The population boom and the increased prices of oil and fertilisers are also to blame. After a period of stability or even decline, the price of wheat, rice, chickpeas, and other staples has gone up. The era of plenty may be near its end. India, long an exporter of food, has begun to import rice. The population explosion in Africa has led to shortage, and although Chinese numbers are under better control, affluence means a shift from cheap grains to expensive beef. World production of meat has gone up fourfold since the 1960s. A carnivorous diet is expensive in many ways. It takes fifty times more energy to make a pound of beef than a pound of corn. The world's fisheries have been depleted, and—in a fatuous gesture of ecological concern—some of its finest land is now used to grow biofuels. As the climate changes, productive regions have lost their worth. Gloom is an occupational hazard for ecologists, but it is getting harder to be cheerful.

The fate of Charles Darwin's experimental subjects is part of a larger litany of decline. The Red List of Threatened Species includes sixteen thousand names, and every year some spectacular creature is declared to be extinct. Many more pass unremarked and unmourned. If weeds and crops are to spread, others must pay the price. Most disappear as their habitats are destroyed. Almost half the rain forest has gone, and mangrove swamps and Mediterranean landscapes face the same disaster with less publicity. Extinction is part of evolution. Around half a million bird species have lived since the group evolved a hundred million years ago, but only about ten thousand lived at any one time—and that is the figure today (although twelve hundred among them are threatened). Even so, it is hard to deny that we live in a time of rapid change, in which some creatures thrive while many more are doomed.

One small and specialised group of mammals—until not long ago a vigorous part of the fauna of Europe, Africa, and Asia—is under particular threat. A world once filled with our hairy relatives will soon be a Planet of the

Ape. Almost half the world's six hundred or so varieties of monkeys and apes face disaster. Habitat destruction is the main threat, although hunters and disease also kill them off. Every large primate except one is close to the end of its evolutionary road. Just two hundred thousand chimps are left in the wild. Gorillas have gone even faster and in some places have died in multitudes from Ebola virus, caught from humans. Many populations of the orangutan, broken up as the forest is destroyed, are already too small to sustain themselves.

The genes show that the great apes have suffered a real reversal over the past several hundred thousand years. The amount of variation in DNA says a lot about the abundance of creatures in ancient times, for populations that stay small for many generations lose genetic variability through the accidents of reproduction, while abundant animals can maintain a pool of diversity that persists for long after any collapse in numbers. The double helix hints that for much of that period orangs, chimps, and gorillas flourished while *Homo sapiens* and his immediate ancestors struggled to survive. Chimps, threatened as they now may be, are three times more different, one from the other, at the molecular level than are humans. Even within the past seventy thousand years, *Homo sapiens* went through a bottleneck of just two or three thousand individuals during a long age of drought (and there have been plenty of local crises since then as humans filled new continents and remote islands). For much of history we were the endangered primate while our relatives boomed. Now the boot is on the other foot.

The world has six billion people (and the figure will rise by half when the population peaks, at around the time of *The Origin*'s bicentennial). *Homo sapiens* has, like the kudzu vine and the earthworm, begun to multiply and to move. By so doing it has revolutionised its own future. Although we may not notice it, our biology is in the middle of a shift as great as that of many of the animals and plants that surround us.

Evolution occurs in space as much as in time. As Darwin saw on the Galapagos—and as is manifest in the geography of human genes—populations isolated from each other diverge, either in response to local forces of natural selection or by random change. Genetic trends have also emerged over the past few thousand years because of accidents of settlement, with reduced levels of variation in recently discovered places such as the New World or the far Pacific. In the same way, ancient patterns arising from the migration of peoples from the Levant to the British Isles have been preserved. Now that history is being lost. We have long been the weediest primate of all but we are becoming weedier still. History has always been made in bed, but the beds are closer together than ever.

Inbreeding, in the developed world at least, is less common than it was. How far was your birthplace from that of your partner, and how far apart were those of your mother and father, and your grandmother and grandfather on each side? For almost everyone the distance has increased over the generations and continues to do so (my wife and I first saw the light three thousand miles apart, my mother and father about three; as my students say, it shows). Even in the Middle East—that centre of sexual conservatism—education, affluence, and the chance to travel mean that DNA is on the move. A series of Israeli Arab villages experienced a drop in the incidence of cousin marriages from about one in four in the 1980s to one in six at the millennium. The same is true in Jordan, Lebanon, and Palestine.

In the past, hurdles in the mind kept people apart. If today's remnants of hunting tribes are any guide, any attempt to join another group was punished by death. European frontiers, too, are marked by genetic steps (with a deeper difference in identity between the beer drinkers of the north and the wine bibbers of Spain and Italy). The continent's long history of staying close to home is manifest in its surnames—those windows into sexual history—as much as its genes. A family tree of names used to fit almost exactly with national boundaries, with the Camerons more or less confined to Scotland and the Zapateros to Spain. Spain itself has the most localised patterns (and can also boast the commonest surname in Europe—Garcia), while in both names and genes Paris is the most diverse city in Europe.

Names—and the DNA that accompanies them—are on the move. In 1881, Darwin's last full year (and the date of a British census), his surname was borne by about one Briton in fifty thousand. Its origin is Welsh, from "derwen," or oak: a name transferred to the River Darwen in Lancashire. In his day its headquarters was in the Sheffield region, where the Darwin tag was eight times more common than in Britain as a whole. Nearly all those who bore it lived within fifty miles of the city (the Joneses—the group with the second most common British surname after Smith—were still, in those happy days, largely confined to Wales, where in some villages they formed a majority). By 1998, the naturalist's name had spread northward to find a new centre in Durham, with secondary nuclei across the north from the Mersey to the Tyne and a minor outbreak in Herefordshire. The Joneses too had migrated, with a great smear across the Welsh borders into northwest England and as far south as London. No longer must a Darwin, a Jones, or anyone else marry someone from their own family for lack of choice. Instead they come into contact with a diversity of potential partners. The proportion of shared names in the marriage records of a typical English village has gone down by 1 or 2 percent a year even since the mid-1970s and by much more since the publication of *The Origin of Species*. Sheffield, once its author's nominal capital, now has scores of new names from across the world.

The United States has gone even further down the road to homogeneity. Its telephone directories contain a million different surnames. Some remnants of history remain, with Wisconsin full of Scandinavians while the phone books of New Mexico, Colorado, and Texas reveal the presence of many Spanish immigrants. The general picture, though, is—unlike Europe's—one of geographic uniformity. One in eight Americans is foreign-born (in California the number is twice as many), and Americans move house, on average, a dozen times during their lifetimes, with ten million each year shifting from one state to another. Such frantic migration soon mixes up names and, in time, genes. That shift—the train or plane as an agent of evolution—is spreading around the globe and will soon even out yet more of man's genetic differences.

Almost everywhere, the biological frontiers are becoming porous. An era of homogeneity is at hand as thousands of people move in search of work or sunshine and, in the end, sex. In Britain, the proportion of those born abroad has doubled in the past fifty years and now represents a tenth of the population. Man, like the ecosystem he lives in, is in the midst of a grand averaging.

Intermarriage has been around for a long time. Many white Britons can trace at least one black ancestor from the small African population that lived in England several centuries ago. About half the men of a certain Yorkshire family, the Revises, share a Y chromosome type otherwise found only in West Africa. There have been Africans in Britain since Roman times, and by the eighteenth century these islands held ten thousand black people. Since then the proportion of Britons who claim recent full or partial descent from Africa has gone up by a factor of twenty. The popular view of the British as a nation walled into a series of ghettos is wrong. In 1991 ethnic minorities in one in ten electoral wards (a ward is the smallest parliamentary subdivision, with around eight thousand wards in England) constituted more than 10 percent of the total citizenry. Ten years later the figure was one in eight, and in 2011 it is expected to be one in five. Most of the growth comes not from immigration but from the movement of people within Britain and from the simple fact that young people, many of them migrants, have more children than do older individuals.

All geneticists are firm believers in the power of lust to overcome social or geographic barriers. In 2001, about one British marriage in fifty—a quarter of a million in all, with many more couples cohabiting—was between partners of different ethnic groups. Hundreds of thousands of children have one parent from Britain and one from the Caribbean, and almost as many are the progeny of white and Asian parents. British Afro-Caribbean males are one and a half times more likely to marry a white woman than black women are to find a white husband (although those preferences are reversed for the Chinese). Such relationships are not, as often believed, found only among the poor, for more than half of those involved live in the suburbs and are richer and more educated than the national average. Assimilation is well under way in modern Britain, one of the most sexually open nations in the world. In today's England, mate choice is influenced as much by level of education as by skin colour. Many other countries, whether they like it or not, are also opening up their gene pools. *Homo sapiens*—already the most geographically tedious of all mammals—will soon, like the worms and the insects, be even more uniform than he was.

In his global coalescence the shaven ape has evolved in much the same fashion as have weedy plants and animals. In other ways, though, man is unique: he is the only animal that has escaped—or almost escaped—Darwin's unforgiving laws of life and death. Natural selection has long been at work on our species, even if our ingenuity has mitigated its power. Now, in the nation in which the idea was invented—and, more and more, in the world as a whole—natural selection has slowed down and may soon stop.

I depress my first-year students with the statement that two out of three of them will die for reasons connected to the genes they carry (a vague claim, but good enough to make my point). Then I try to cheer them up by pointing out that had I given the lecture in Shakespeare's time, two out of three of them would, at age eighteen or so, be dead already. Even in the year of Darwin's birth, about half of all British newborns died before they reached maturity. Life has seen a great change for the better. An English baby born in the year of the millennium had a 99 percent chance of surviving until 2021, and that figure continues to improve. Japan does even better, and the United States rather worse, but most of the developed world has seen a revolution in the prospects of the young. In most countries, even relatively poor ones like

Ecuador, the death rate of children is no more than twice the British figure. In some places it is less. Africa is, alas, a great exception. The death rate for children under five is one in four in Sierra Leone and almost as high in Angola and Liberia, while Swaziland has an overall life expectancy of just forty years, half that of Japan. Differences in childhood mortality thus still account for most of the world's variation in life expectancy, but outside the continent where our species began, those differences have withered away.

Natural selection exerts much of its power through the death of young people, for they have not yet passed on their genes. The death of the elderly—those over forty or so—is a less potent agency, for their sexual lives are more or less over and their relevance to evolution at an end. As a result, the great agent of change has lost much of its fuel. For most of history, the Grim Reaper was the master of man's biological fate. Now he is taking a rest, and inherited differences in the ability to withstand cold, starvation, vitamin deficiency, or infectious disease no longer drive the evolutionary machine. Plenty of people still die for those reasons, but they do so after selection loses interest in them.

The Darwinian examination has two parts. In the developed world, most people pass the first: they stay alive until they are old enough to have children. The second section is just as inexorable but has a wider range of marks: any candidate for evolutionary success has to find a mate and reproduce. The more children they have, the better the prospects for their genes.

Among mammals, females are limited in the number of offspring they can produce by the mechanics of pregnancy and child care, while males are free to spread their sperm to many partners (although a certain amount of persuasion may be needed). As a result, males compete for the attention of females, while females must decide which males should be allowed in. Sexual selection turns on the same logic as natural selection: it depends on inherited differences, not in the chances of life or death but in the number of young. The rule applies as much to humans as it does to birds and flowers.

Both humans and peacocks have more variation in male sexual success than in female. Until not long ago, many societies contained a few satisfied libertines, outnumbered by a mass of sexually frustrated men. The powerful have always taken their amatory chances when they arise. Henry VIII was a minor player in the marital stakes. The Emperor in the Ch'I dynasty of China maintained a palace with several thousand women available for his amusement, while in tribal societies men with high status still have many more mates than their lowly (and often celibate) fellows. Mohamed bin Laden, father of Osama, had twenty-two wives and fifty-three children (and in the year of Osama's birth he had six). His best-known son had, last time they were counted, five wives and twenty-two children. Given the equal numbers of men and women at birth, plenty of his henchmen will be obliged to die, naturally or otherwise, entirely childless.

The British Isles themselves have a history of sexual inequality that makes the antics of the founder of the Church of England look feeble. The evidence is in the Y chromosome, which marks male descent. It comes in a great diversity of form. In most places most men have their own more or less unique model. A fifth of the men of northwest Ireland, in contrast, share a more or less identical version of the Y chromosome—which means that they all trace ancestry from the same male.

Old Erin was rife with sexual inequality. Lord Turlough O'Donnell, who died in 1423, had eighteen sons and fifty-nine grandsons. He was himself a descendant of the High Kings of Ireland, all of whom claimed a single fifth century warlord, Niall of the Nine Hostages, as their shared ancestor. The Y chromosome of Niall the hostage taker has, thanks to his exploits and those of his descendants, spread to one in twelve of today's Irishmen in Ireland and to millions more across the globe. The surnames fit too, for men of the Gallagher, O'Neill, and Quinn families (all of whom claim descent from the High Kings) are most likely to bear the special Y chromosome. In Ireland, for many centuries, the mightiest male passed on his genes, and many of his fellows passed their days in glum celibacy, occupying themselves as soldiers or priests instead.

Today's reproductive universe is quite different. Everywhere the weak and the powerful—the poor and the rich—are sexually closer than they were. In almost every social class, the average number of children has, thanks to technology, gone down. Natural selection cares naught for that, for the important figure is not the number of offspring but the variation in how many progeny people have. That figure has shrunk. Five centuries ago in Florence, the upper crust had twice as many offspring as did the peasantry, but now the Florentine poor have slightly more than the rich; Britain is the same (which worries the right-wing press). The gulf has closed through restraint by the affluent rather than excess by the poor. A furtive exchange of contraceptive information meant that rich families became smaller while those of the poor stayed the same but now both groups are able to control the number of children they have. Schools too are a powerful agent of birth control. Everywhere, people with degrees have fewer offspring than those who drop out early. As education spreads, the fertility imbalances will become smaller still.

Inequality, in survival or in sex, is the raw material of evolution. The differences in people's ability to stay alive and have children can be combined in a single measure that shows just how much of it is still available to evolution. Around the world, that figure—the "opportunity for selection"—is in decline. India tells the tale within a single country. The nation encompasses a range of cultures, from tribal hill peoples to affluent urbanites, together with vast numbers of peasant farmers whose lives resemble those of Europeans a few centuries ago. The figures of life, death, and birth, when put together, show that natural selection has lost nine-tenths of its power in India's middle class when compared with its tribal peoples. Much the same is true when we compare the modern world with that of the Middle Ages and—to a lesser degree—even with that of the Victorians.

As Charles Darwin insisted, evolution is not a predictive science. Natural selection has no inbuilt tendency to improve matters (or to make them worse). For *Homo sapiens*, some nasty surprises no doubt lurk around the corner. One day, his machine will take its revenge, and we may well fail in the struggle for existence against ourselves, the biggest ecological challenge of all.

Whatever the future holds, two centuries after Darwin's birth we have entered a new era in biology. The changes are not limited to the rain forest, the coral reefs, or the teeming tropics but are hard at work in Charles Darwin's homeland and throughout the world. From Shrewsbury to the Galapagos, and from worms to barnacles to human beings, there has been a triumph of the average. The Earth as an island rather than an archipelago is a far less interesting place than when HMS *Beagle* set sail. Whether it becomes even less so—and whether it survives at all—depends on the talents of the only creature ever to step beyond the limits of Darwin's theory and to be capable of concern about the future.

Annotated Bibliography

PREFACE

Browne, Janet. 1996. *Charles Darwin: A biography, volume 1—Voyaging*. Princeton, NJ: Princeton University Press.

Browne, Janet. 2003. *Charles Darwin: A biography, volume 2—The power of place*. Princeton, NJ: Princeton University Press.

The canonical biography of Charles Darwin: The story of his life and the formation of his scientific ideas. Volume 2 concentrates on his work after the return from the *Beagle* voyage.

Ghiselin, Michael T. 2003. *The triumph of the Darwinian method*. Mineola, NY: Dover Publications.

Demonstrates the unity of thought that pervades Darwin's books, from first to last.

CHAPTER 1: THE QUEEN'S ORANG-UTAN

Allison, A. C. 2009. Genetic control of resistance to human malaria. *Current Opinion in Immunology* 21:499–505.

The complexity of the evolved response to a single disease.

Bakewell, M. A., Shi, P., and Zhang, J. 2007. More genes underwent positive selection in chimpanzee evolution than in human evolution. *Proceedings of the National Academy of Sciences, USA* 104:7489–7494.

Slower evolution in the human than in the chimpanzee line.

Fisher, S. E., and Scharff, C. 2009. FOXP2 as a molecular window into speech and language. *Trends in Genetics* 25:166–177.

Mutations in a developmental gene as possible key to the origin of language.

Gibbons, A. 2009. *Ardipithecus* unveiled. *Science* 326:36–40.

Introduction to a series of linked papers on the anatomy, ecology, and behaviour of a 4.4-million-year-old human ancestor.

Gibbs, R. A., Rogers, J., Katze, M. G., et al. 2007. Evolutionary and biomedical insights from the rhesus macaque genome. *Science* 316:222–234.

Some surprising differences between us and a moderately close relative.

Go, Y., and Niimura, Y. 2008. Similar numbers but different repertoires of olfactory receptor genes in humans and chimpanzees. *Molecular Biology and Evolution* 25:1897–1907.

Decay and decline in the human compared with the chimpanzee line.

Herrmann, E., Call, J., Hernández-Lloreda, M. V., Hare, B., and Tomasello, T. 2007. Humans have evolved specialized skills of social cognition: The cultural intelligence hypothesis. *Science* 317:1360–1365.

Homo sapiens as more than a shaved monkey.

Ingram, C. J. E., Mulcare, C. A., Itan, Y., Thomas, M. G., and Swallow, D. M. 2009. Lactose digestion and the evolutionary genetics of lactase persistence. *Human Genetics* 124:579–591.

Evolution of milk-drinking in cattle-herding societies.

Lao, O., de Gruijter, J. M., van Duijn, K., Navarro, A., and Kayser, M. 2007. Signatures of positive selection in genes associated with human skin pigmentation as revealed from analyses of single nucleotide polymorphisms. *Annals of Human Genetics* 71:354–369.

Indirect evidence of natural selection on the human genome.

Li, J. Z., Absher, D. M., Tang, H., Southwick, A. M., Casto, A. M., Ramachandran, S., Cann, H. M., Barsh, G. S., Feldman, M., Cavalli-Sforza, L. L., and Myers, R. M. 2008. Worldwide human relationships inferred from genome-wide patterns of variation. *Science* 319:1100–1104.

The human genome diversity project and overall levels of human genetic variation.

Marques-Bonet, T., Ryder, O. A., and Eichler, E. E. 2009. Sequencing primate genomes: What have we learned? *Annual Review of Genomics and Human Genetics* 10:355–386.

Ourselves and our relatives compared.

Reed, D. L., Light, J. E., Allen, J. M., and Kirchman, J. J. 2007. Pair of lice lost or parasites regained: The evolutionary history of anthropoid primate lice. *BioMedCentral Biology* 5:7.

The intimate secrets of body hair and the invention of clothes.

Rossiianov, K. 2002. Beyond species: Il'ya Ivanov and his experiments on cross-breeding humans with anthropoid apes. *Science in Context* 15:277–316.

A failed attempt to use biology to ask what makes us human.

Sturm, R. A. 2009. Molecular genetics of human pigmentation diversity. *Human Molecular Genetics* 18(R1):R9–R17.

Genetics and evolution of skin, hair, and eye colour.

Tattersall, I., and Schwartz, J. H. 2009. Evolution of the genus *Homo. Annual Review of Earth and Planetary Sciences* 37:67–92.

The tangled tale of our relatively recent ancestors and supposed ancestors.

Varki, A., Geschwind, D. H., and Eichler, E. E. 2008. Explaining human uniqueness:

Genome interactions with environment, behaviour, and culture. *Nature Reviews Genetics* 9:749–763.
What makes us different from apes, in brain and body.

CHAPTER 2: THE GREEN TYRANNOSAURS

Bauer, U., Bohn, H. F., and Federle, W. 2008. Harmless nectar source or deadly trap: *Nepenthes* pitchers are activated by rain, condensation, and nectar. *Proceedings of the Royal Society of London, Series B* 275:259–265.
Trapping strategies of pitcher plants.

Clement, L. W., Köppen, C. W., Brand, W. A., and Heil, M. 2008. Strategies of a parasite of the ant-Acacia mutualism. *Behavioral Ecology and Sociobiology* 26:953–962.
Fine balance of mutualism versus parasitism in the ant-tree interaction.

Eilenberg, E., Pnini-Cohen, S., Schuster, S., Movtchan, A., and Zilberstein, A. 2006. Isolation and characterization of chitinase genes from pitchers of the carnivorous plant *Nepenthes khasiana*. *Journal of Experimental Botany* 57:2775–2784.
How botanical carnivores digest their prey.

Ellison, A. M., and Gotelli, N. J. 2009. Energetics and the evolution of carnivorous plants—Darwin's "most wonderful plants in the world." *Journal of Experimental Botany* 60:19–42.
The costs and benefits of eating insects.

Forterre, Y., Skotheim, J. M., Dumais, J., and Mahadevan, L. 2005. How the Venus fly-trap snaps. *Nature* 433:421–425.
Biophysics of the wonderful plant.

Frederickson, M. E., Greene, M. J., and Gordon, D. M. 2005. 'Devil's gardens' bedevilled by ants. *Nature* 437:495–496.
How mutualist ants on trees destroy neighbouring vegetation.

Gibson, T. C., and Waller, D. M. 2009. Evolving Darwin's 'most wonderful' plant: Ecological steps to a snap-trap. *New Phytologist* 183:575–587.
The evolution of a most unlikely structure, the Venus flytrap.

Gorb, E., Haas, K., Henrich, A., Enders, S., Barbakadze, N., and Gorb, S. 2005. Composite structure of the crystalline epicuticular wax layer of the slippery zone in the pitchers of the carnivorous plant *Nepenthes alata* and its effect on insect attachment. *Journal of Experimental Biology* 208:4651–4662.
The complexity of the pitfall in a pitcher plant.

Heil, M., and McKey, D. 2003. Protective ant-plant interactions as model systems in ecological and evolutionary research. *Annual Review of Ecology, Evolution, and Systematics* 34:425–453.
Review of the mutualist and agonistic interactions between ants and trees.

Juniper, B. E., Robins, R. J., and Joel, D. M. 1989. *The carnivorous plants*. London: Academic.
Botany, anatomy, and taxonomy of many species of insect-eaters.

Karagatzides, J. D., and Ellison, A. M. 2009. Construction costs, payback times, and the leaf economics of carnivorous plants. *American Journal of Botany* 96:1612–1619.
When to eat insects, and when to give it up.

Murza, G. L., Heaver, J. R., and Davis, A. R. 2006. Minor pollinator-prey conflict in the carnivorous plant *Drosera anglica*. *Plant Ecology* 184:43–52.
Divergence of interest between insect and plant: food item or sexual aid?
Palmer, T. M., and Brody, A. K. 2007. Mutualism as reciprocal exploitation: African plant-ants defend foliar but not reproductive structures. *Ecology* 88:3004–3011.
Fine balance of advantage between ants and trees.
Raaijmakers, J. M., Paulitz, T. C., Steinberg, C., Alabouvette, C., and Moënne-Loccoz, Y. 2009. The rhizosphere: A playground and battlefield for soil-borne pathogens and beneficial microorganisms. *Plant and Soil* 321:341–361.
Complexity of interaction between plants and nitrogen-fixing insects.

CHAPTER 3: SHOCK AND AWE

Blasi, G., Hariri, A. R., Alce, G., Taurisano, P., Sambataro, F., Das, S., Bertolino, A., Weinberger, D. R., and Mattay, V. S. 2009. Preferential amygdala reactivity to the negative assessment of neutral faces. *Biological Psychiatry* 66:847–853.
Brain scans and response to subjective feelings when looking at faces.
Canli, T., Ferri, J., and Duman, E. A. 2009. Genetics of emotion regulation. *Neuroscience* 164:43–54.
Individual variation in emotional response within the normal range.
Davidson, R. J. 2003. Darwin and the neural bases of emotion and affective style. *Annals of the New York Academy of Sciences* 1000:316–336.
The Expression of the Emotions put into twenty-first century context.
Frith, C. D., and Frith, U. 2007. Social cognition in humans. *Current Biology* 17:R724–R732.
Importance of understanding messages of mood and intention for society.
Gaspar, A. 2006. Universals and individuality in facial behavior—past and future of an evolutionary perspective. *Acta Ethologica* 9:1–14.
Parallels between the facial expressions of primates and humans.
Haberman, J., and Whitney, D. 2007. Rapid extraction of mean emotion and gender from sets of faces. *Current Biology* 17:R751–R753.
Ability to sense average emotions of a group of people from a mere glimpse.
Hare, B. 2007. From nonhuman to human mind: What changed and why? *Current Directions in Psychological Science* 16:60–66.
Social skills and emotional response in dogs, apes, and humans compared.
Hariri, A. R., and Holmes, A. 2006. Genetics of emotional regulation: The role of the serotonin transporter in neural function. *Trends in Cognitive Sciences* 10:182–186.
Individual differences in efficiency of nerve transmission alter anxiety and mood.
Kanwisher, N., and Yovel, G. 2006. The fusiform face area: A cortical region specialized for the perception of faces. *Philosophical Transactions of the Royal Society of London, Series B* 361:2109–2128.
Brain area lights up in specific response to face; effects of age, race, and experience.
Kendrick, K. M. 2006. The neurobiology of social recognition, attraction, and bonding. *Philosophical Transactions of the Royal Society of London, Series B* 361:2057–2059.
Nervous and hormonal correlates of response to familiar and unfamiliar faces.
Levy, S. E., Mandell, D. S., and Schultz, R. T. 2009. Autism. *The Lancet* 374:1627–1638.

Review of symptoms, incidence, and genetics of autism and related disorders.
Miklosi, A. 2007. *Dog Behaviour, Evolution, and Cognition*. New York: Oxford University Press.
Social skills of dogs: response to owner versus strangers, ability to sense mood.
Mitani, J. C. 2009. Cooperation and competition in chimpanzees: Current understanding and future challenges. *Evolutionary Anthropology* 18:215–227.
The emotional universe of the chimpanzee: anger and suspicion versus cooperation.
Niedenthal, P. M. 2007. Embodying emotion. *Science* 316:1002–1009.
Emotion as told in body language and facial expression.
Palermo, R., and Rhodes, G. 2007. Are you always on my mind? A review of how face perception and attention interact. *Neuropsychologia* 45:75–92.
Familiarity of faces and speed of perception; disorders of face recognition.
Phelps, E. A. 2006. Emotion and cognition: Insights from studies of the human amygdala. *Annual Review of Psychology* 57:27–53.
The base of the brain and response to fearful or angry expressions.
Russell, J. A., Bachorowski, J. A., and Fernández-Dols, J. M. 2003. Facial and vocal expressions of emotion. *Annual Review of Psychology* 54:329–349.
Ambiguity of expressions such as laughter; balance between signal and reception.
Spady, T. C., and Ostrander, E. A. 2008. Canine behavioral genetics: Pointing out the phenotypes and herding up the genes. *American Journal of Human Genetics* 82: 10–18.
Crosses between breeds and the genetics of behavioural and emotional differences.
Sugita, Y. 2009. Innate face processing. *Current Opinion in Neurobiology* 19:39–44.
Innate ability to see faces in very young primates.
Tate, A. J., Fischer, H., Leigh, A. E., and Kendrick, K. M. 2006. Behavioural and neurophysiological evidence for face identity and face emotion processing in animals. *Philosophical Transactions of the Royal Society of London, Series B* 361:2155–2172.
Apes and even sheep able to recognize emotions shown on faces.

CHAPTER 4: THE TRIUMPH OF THE WELL BRED

Barrett, S. C. H. 2003. Mating strategies in flowering plants: The outcrossing-selfing paradigm and beyond. *Philosophical Transactions of the Royal Society of London, Series B* 358:991–1004.
More flexibility in mating strategy by plant species than imagined by Darwin.
Bittles, A. H. 2008. A community genetics perspective on consanguineous marriage. *Community Genetics* 11:324–330.
Incidence and health effects of inbreeding and effects on clinical genetics services.
Charlesworth, D., and Willis, J. H. 2009. The genetics of inbreeding depression. *Nature Reviews Genetics* 10:783–796.
How inbreeding reduces diversity and exposes hidden genetic damage.
Colantonio, S. E., Lasker, G. W., Kaplan, B. A., and Fuster, V. 2003. Use of surname models in human population biology: A review of recent developments. *Human Biology* 75:785–807.
Isonymy—marriage of those with shared surnames—as an indication of inbreeding.
Helgason, A., Pálsson, S., Gudbjartsson, D., Kristjánsson, D., and Stefánsson, K. 2008.

An association between the kinship and fertility of human couples. *Science* 319:813–816.

Unexpected reduction in fertility in later generations after cousin marriage.

Ilmonen, P., Stundner, G., Thoss, M., and Penn, D. J. 2009. Females prefer the scent of outbred males: Good-genes-as-heterozygosity? *BioMedCentral Evolutionary Biology* 9:104.

Female mice can assess inbreeding of males by scent and avoid inbreds.

Jorde, L. 2001. Consanguinity and prereproductive mortality in the Utah Mormon population. *Human Heredity* 52:61–65.

Consistency of effects of cousin marriage over a century in a well-studied group.

Keller, L. F., and Waller, D. M. 2002. Inbreeding effects in wild populations. *Trends in Ecology and Evolution* 17:230–241.

Similar reductions in fitness of inbred offspring in birds, mammals, and plants.

King, T. E., Ballereau, S. J., Schürer, K. E., and Jobling, M. A. 2006. Genetic signatures of coancestry within surnames. *Current Biology* 16:384–388.

Shared rare surnames mean shared genes; proof of Darwin's ideas on isonymy.

Lieberman, D., Tooby, J., and Cosmides, L. 2007. The architecture of human kin detection. *Nature* 445:727–732.

Older children less sexually interested in their younger siblings than vice versa.

Milinski, M. 2006. The major histocompatibility complex, sexual selection, and mate choice. *Annual Review of Ecology, Evolution, and Systematics* 37:159–186.

Outbreeding increases immunological diversity and disease resistance: mate choice.

Owens, S. J., and Miller, R. 2009. Cross- and self-fertilization of plants—Darwin's experiments and what we know now. *Botanical Journal of the Linnean Society* 161:357–395.

Darwin's crossing experiments reinterpreted in the light of modern science.

Pattison, J. E. 2004. A comparison of inbreeding rates in India, Japan, Europe, and China. *HOMO—Journal of Comparative Human Biology* 55:113–128.

Historical inbreeding influenced by political turmoil and movement.

Rudan, I., Campbell, H., Carothers, A. D., Hastie, N. D., and Wright, A. F. 2006. Contribution of consanguinity to polygenic and multifactorial diseases. *Nature Genetics* 11:1224–1225.

Inbred offspring are more susceptible to a wide range of diseases.

Snowdon, C. T., Ziegler, T. E., Schultz-Darken, N. J., and Ferris, C. F. 2006. Social odours, sexual arousal, and pairbonding in primates. *Philosophical Transactions of the Royal Society of London, Series B* 361:2079–2089.

Assessment of kinship by scent and inbreeding avoidance in apes.

Weller, S. G. 2009. The different forms of flowers—What have we learned since Darwin? *Botanical Journal of the Linnean Society* 160:249–261.

How primrose pins and thrums grew into today's science of sexual choice.

CHAPTER 5: THE DOMESTIC APE

Bell, C. G., Walley, A. J., and Froguel, P. 2005. The genetics of human obesity. *Nature Reviews Genetics* 6:221–227.

The cost of abundant food to the most domesticated primate.

Brown, T. A., Jones, M. K., Powell, W., and Allerby, R. G. 2008. The complex origins of

domesticated crops in the Fertile Crescent. *Trends in Ecology and Evolution* 24:103–109.
Domestication as a more protracted and diffuse process than once thought.
Doebley, J. F., Gaut, B. S., and Smith, B. D. 2006. The molecular genetics of crop domestication. *Cell* 127:1309–1321.
The maize genome and the great reorganisation involved in domestication.
Gotherstrom, A., Anderung, C., Hellborg, L., Elburg, R., Smith, C., Bradley, D. G., and Ellegren, H. 2005. Cattle domestication in the Near East was followed by hybridization with aurochs bulls in Europe. *Proceedings of the Royal Society of London, Series B* 272:2345–2350.
Continued crossing of farm cows with wild bulls during early cattle domestication.
Lia, V. V., Confalonieri, V. A., Ratto, N., Hernandez, J. A. C., Alzogaray, A. M. M., Poggio, L., and Brown, T. A. 2007. Microsatellite typing of ancient maize: Insights into the history of agriculture in southern South America. *Proceedings of the Royal Society of London, Series B* 274:545–554.
Tracking the geography of domestication of maize in South America.
Ostrander, O. A., and Wayne, R. K. 2005. The canine genome. *Genome Research* 15:1706–1716.
The genetics of the most domesticated of all animals.
Parker, H. G., Kim, L. V., Sutter, N. B., Carlson, S., Lorentzen, T. D., Malek, T. B., Johnson, G. S., DeFrance, H. B., Ostrander, E. A., and Kruglyak, L. 2004. Genetic structure of the purebred domestic dog. *Science* 304:1160–1164.
Extensive genetic divergence among dog breeds after only a few generations.
Purugganan, M. D., and Fuller, D. Q. 2009. The nature of selection during plant domestication. *Nature* 457:843–848.
Parallel changes in seed size, shattering ability, and growth habit in various crops.
Saetre, P. 2004. From wild wolf to domestic dog: Gene expression changes in the brain. *Molecular Brain Research* 126:198–206.
Dogs have very different patterns of gene activity in the brain following domestication.
Sampietro, M. L., Lao, O., Caramelli, D., Lari, M., Pou, R., Marti, R., Bertranpetit, J., and Lalueza-Fox, C. 2007. Palaeogenetic evidence supports a dual model of Neolithic spreading into Europe. *Proceedings of the Royal Society of London, Series B* 274:2161–2167.
Complexity of interaction of early farmers with European hunter-gatherers.
Trut, L., Oskina, I., and Kharlamova, A. 2009. Animal evolution during domestication: The domesticated fox as a model. *BioEssays* 31:349–360.
Changes in behaviour, appearance, and gene expression in domesticated foxes.
Vaughan, D. A., Balazs, E., and Heslop-Harrison, J. S. 2007. From crop domestication to super-domestication. *Annals of Botany* 100:893–901.
Origin of maize and other crops; potential for genetic manipulation in future.

CHAPTER 6: THE THINKING PLANT

Baldwin, I. T. 2006. Volatile signaling in plant-plant interactions: "Talking trees" in the genomics era. *Science* 311:812–816.
Chemical communication by scent among plants after damage.

Braam, J. 2005. In touch: Plant responses to mechanical stimuli. *New Phytologist* 165:373–389.
The touch genes in plants: effect on growth form, movement of leaves, and more.

Chen, M., Chory, J., and Fankhauser, C. 2004. Light signal transduction in higher plants. *Annual Review of Genetics* 38:87–117.
The complexity of response by plants to colour, intensity, and timing of light.

Franklin, K. A., and Whitelam, G. C. 2005. Phytochromes and shade-avoidance responses in plants. *Annals of Botany* 96:169–175.
How shaded plants struggle towards the light.

Gianoli, E. 2004. Evolution of a climbing habit promotes diversification in flowering plants. *Proceedings of the Royal Society of London, Series B* 271:2011–2015.
Climbers find a new way of life and explode into variety.

Goriely, A., and Neukirch, S. 2006. Mechanics of climbing and attachment in twining plants. *Physical Review Letters* 97, 184302.
The laws of physics determine how climbing plants conquer their supporters.

Harsh, P., Bais, T. L., Weir, L. G., Perry, S. G., and Vivanco, J. M. 2006. The role of root exudates in rhizosphere interactions with plants and other organisms. *Annual Review of Plant Biology* 57:233–266.
Plant roots hunting for nitrogen and other nutriments.

Karban, R. 2008. Plant behaviour and communication. *Ecology Letters* 11:727–739.
Parallels between plant and animal foraging, defense, and communication.

Kiss, J. Z. 2006. Up, down, and all around: How plants sense and respond to environmental stimuli. *Proceedings of the National Academy of Sciences, USA* 103:829–830.
Using mutants to disentangle the sensory world of plants.

Krings, M., Kerp, H., Taylor, T. N., and Taylor, E. L. 2003. How Paleozoic vines and lianas got off the ground: On scrambling and climbing Carboniferous–early Permian pteridosperms. *Botanical Review* 69:204–224.
Incidence of the climbing habit in coal-measure forests.

Palmieri, M., and Kiss, J. Z. 2006. The role of plastids in gravitropism. *Advances in Photosynthesis and Respiration* 23:507–525.
Parallel between the mammalian ear and gravity sensing organs in plants.

Quint, M., and Gray, W. M. 2006. Auxin signalling. *Current Opinion in Plant Biology* 9:448–453.
Review of the plant hormone auxin and its role in movement and growth.

Runyon, J. B., Mescher, M. C., and De Moraes, C. M. 2006. Volatile chemical cues guide host location and host selection by parasitic plants. *Science* 313:1964–1967.
Ability of dodder plants to sniff out their lush prey and avoid less desirable victims.

Santner, A., and Estelle, M. 2009. Recent advances and emerging trends in plant hormone signalling. *Nature* 459:1071–1078.
The molecular biology of auxins and other plant hormones.

Schnitzer, S. A., and Bongers, F. 2002. The ecology of lianas and their role in forests. *Trends in Ecology and Evolution* 17:223–231.
Biomass of lianas in forests; ability to exploit light patches and damage support.

Telewski, F. W. 2006. A unified hypothesis of mechanoperception. *American Journal of Botany* 93:1466–1476.
A sensory network within plant cells acts as a transducer for touch perception.

Trewavas, A. 2003. Aspects of plant intelligence. *Annals of Botany* 92:1–20.
The case for plants having a mental capacity analogous to that of animals.

Vandenbussche, F., Pierik, R., Millenaar, F. F., Voesenek, L. A. C. J., and van der Straeten, D. 2005. Reaching out of the shade. *Current Opinion in Plant Biology* 8:462–468.
Plants use balance of red and infrared light to sense shading by other leaves.

Whippo, C. W., and Hangarter, R. P. 2009. The "sensational" power of movement in plants: A Darwinian system for studying the evolution of behavior. *American Journal of Botany* 96:2115–2127.
Darwin and plant movement: his experiments seen in the context of today's science.

CHAPTER 7: A PERFECT FOWL

Anderson, D. T. 1994. *Barnacles: Structure, function, development and evolution.* London: Chapman and Hall.
The standard work on the taxonomy, anatomy, and distribution of barnacles.

Angelini, D. R., and Kaufman, T. C. 2005. Comparative developmental genetics and the evolution of arthropod body plans. *Annual Review of Genetics* 39:95–119.
Homeoboxes and the evolution of segments in insects, crabs, and barnacles.

Christensen-Dalsgaard, J., and Carr, C. E. 2008. Evolution of a sensory novelty: Tympanic ears and the associated neural processing. *Brain Research Bulletin* 75:365–370.
How marine wave-detecting cells were hijacked for use in the mammalian inner ear.

Damen, W. G. M. 2007. Evolutionary conservation and divergence of the segmentation process in arthropods. *Developmental Dynamics* 236:1379–1391.
Comparison of insect and marine developmental genes reveals deep similarity.

Deutsch, J. S., and Mouchel-Vielh, E. 2003. Hox genes and the crustacean body plan. *BioEssays* 25:878–887.
Variation in the numbers of legs and segments depends on homeobox evolution.

Géant, E., Mouchel-Vielh, E., Coutanceau, J. P., Ozouf-Costaz, C., and Deutsch, J. S. 2006. Are cirripedia hopeful monsters? Cytogenetic approach and evidence for a Hox gene cluster in the cirripede crustacean *Sacculina carcini. Development, Genes, and Evolution* 216:443–449.
How loss of a set of developmental genes allowed barnacles to diversify.

Glenner, H., and Hebsgaard, M. B. 2006. Phylogeny and evolution of life history strategies of the parasitic barnacles (Crustacea, Cirripedia, Rhizocephala). *Molecular Phylogenetics and Evolution* 41:528–538.
Evolution of the macabre lifestyle of the barnacle parasites of crabs.

Lemons, D., and McGinnis, W. 2006. Genomic evolution of Hox gene clusters. *Science* 313:1918–1923.
The Hox developmental genes put into the mammalian and insect context.

Perez-Losada, M., Harp, M., Hoeg, J. T., Achituv, Y., Jones, D., Watanabe, H., and Crandall, K. A. 2007. The tempo and mode of barnacle evolution. *Molecular Phylogenetics and Evolution* 46:328–346.
Molecular trees in barnacles show hidden complexity and convergent evolution.

Shubin, N. 2008. *Your inner fish: A journey into the 3.5-billion-year history of the human body.* New York: Pantheon.
The imprint of our marine past in our own body plan.

Vonk, F. J., and Richardson, M. K. 2008. Developmental biology: Serpent clocks tick faster. *Nature* 454:282–283.
How snakes multiply their somites to generate many vertebrae.

CHAPTER 8: WHERE THE BEE SNIFFS

Anderson, B., Johnson, S. D., and Carbutt, C. 2005. Exploitation of a specialized mutualism by a deceptive orchid. *American Journal of Botany* 92:1342–1349.
How a nectar-free mimic in South Africa parasitizes its more generous neighbours.

Arditti, J. 1992. *Fundamentals of Orchid Biology.* New York: John Wiley and Sons.
Exhaustive account of orchid taxonomy and pollination mechanisms.

Bascompte, J. 2007. Plant-animal mutualistic networks: The architecture of biodiversity. *Annual Review of Ecology, Evolution, and Systematics* 38:567–593.
The ancient and conflict-ridden tie between insects and plants, orchids included.

Chittka, L., and Raine, N. E. 2006. Recognition of flowers by pollinators. *Current Opinion in Plant Biology* 9:428–435.
How the insect visual system is attuned to—and can be fooled by—floral signals.

Fenster, C. B., Armbruster, W. S., Wilson, P., Dudash, M. R., and Thomson, J. D. 2004. Pollination syndromes and floral specialization. *Annual Review of Ecology, Evolution, and Systematics* 35:375–403.
Evolution of shared flower specialisations in unrelated groups of plants.

Friis, E. M., Pedersen, K. R., and Crane, P. R. 2005. When Earth started blooming: Insights from the fossil record. *Current Opinion in Plant Biology* 8:5–12.
The origin of flowers and the evolutionary explosion of blooms and pollinators.

Galliot, C., Stuurman, J., and Kuhlemeier, C. 2006. The genetic dissection of floral pollination syndromes. *Current Opinion in Plant Biology* 9:78–82.
Pollinator shifts as a force driving the origin of new species of plants.

Jersakova, J., Johnson, S. D., and Kindlmann, P. 2006. Mechanisms and evolution of deceptive pollination in orchids. *Biological Reviews of the Cambridge Philosophical Society* 81:219–235.
The infinite dishonesty of orchids in search of a cheap pollinator.

Raguso, R. A. 2004. Flowers as sensory billboards: Progress towards an integrated understanding of floral advertisement. *Current Opinion in Plant Biology* 7:434–440.
The complexity of floral signals from colour to scent; honesty and dishonesty.

Ramírez, S. R., Gravendeel, B., Singer, R. B., Marshall, C. R., and Pierce, N. E. 2007. Dating the origin of the Orchidaceae from a fossil orchid with its pollinator. *Nature* 448:1042–1045.
An ancient pollinator preserved in amber pushes back the early history of orchids.

Specht, C. D., and Bartlett, M. E. 2009. Flower evolution: The origin and subsequent diversification of the angiosperm flower. *Annual Review of Ecology, Evolution, and Systematics* 40:217–243.
Pollinator conflict as a driving force in the evolution of modern flowers.

Wade, M. J. 2007. The co-evolutionary genetics of ecological communities. *Nature Reviews Genetics* 8:185–195.
Pervasive conflicts between pollinators and flowers, hosts and parasites, and more.

Whittall, J. B., and Hodges, S. A. 2007. Pollinator shifts drive increasingly long nectar spurs in columbine flowers. *Nature* 447:706–710.
An evolutionary race between host and pollinator drives both to extremes.
Yam, T. W., Arditti, J., and Cameron, K. M. 2009. "The orchids have been a splendid sport"—An alternative look at Charles Darwin's contribution to orchid biology. *American Journal of Botany* 96:2128–2154.
How Darwin's ideas underpin the latest research in orchid biology.

CHAPTER 9: THE WORMS CRAWL IN

Ashbee, P., and Jewell, P. 1998. The experimental earthworks revisited. *Antiquity* 72:485–504.
Artificial burial mound on downland rapidly disturbed by earthworms.
Bardgett, R. D. 2005. *The biology of soil: A community and ecosystem approach.* Oxford: Oxford University Press.
Review of earthworms as ecological engineers in a complex ecosystem.
Briones, M. J. I., Ostle, N. J., and Pearce, T. G. 2008. Stable isotopes reveal that the calciferous gland of earthworms is a CO_2-fixing organ. *Soil Biology and Biochemistry* 40:554–557.
Ability of worms to extract carbon from air extends role in soil improvement.
Brown, G. G., Feller, C., Blanchart, E., Deleporte, P., and Chernyanskii, S. S. 2003. With Darwin, earthworms turn intelligent and become human friends. *Pedobiologia* 47:924–933.
Changes in attitudes towards earthworms before and after Darwin's day.
Canti, M. G. 2003. Earthworm activity and archaeological stratigraphy: A review of products and processes. *Journal of Archaeological Science* 30:135–148.
How worm activity disturbs archaeological sites and blurs the record of history.
Edwards, C. A., ed. 2004. *Earthworm ecology.* 2nd ed. Boca Raton, FL: CRC Press.
Exhaustive review of ecology, physiology, behaviour, and distribution of worms.
Evans, C. 2009. Small agencies and great consequences: Darwin's archaeology. *Antiquity* 83:475–488.
How archaeological digs showed the power of earthworms to inter.
Gabet, E. J., Reichman, O. J., and Seabloom, E. W. 2003. The effects of bioturbation on soil processes and sediment transport. *Annual Review of Earth and Planetary Sciences* 31:249–273.
Mites, mice, elephants, roots, and more assist earthworms in disturbing the soil.
Glaser, B. 2006. Prehistorically modified soils of central Amazonia: A model for sustainable agriculture in the twenty-first century. *Philosophical Transactions of the Royal Society of London, Series B* 362:187–196.
The terra preta soils of South America as a model for use of biochar in farming.
Johnson, D. L., Domier, J. E. J., and Johnson, D. N. 2005. Reflections on the nature of soil and its biomantle. *Annals of the Association of American Geographers* 95:11–31.
The relative roles of chemistry and biology in making and improving soil.
Meysman, F. J. R., Middelburg, J. J., and Heip, C. H. R. 2006. Bioturbation: A fresh look at Darwin's last idea. *Trends in Ecology and Evolution* 21:688–695.
How disturbing the surface led to the Cambrian Explosion of life.

Montgomery, D. R. 2007. *Dirt: The erosion of civilizations.* Berkeley: University of California Press.
How neglect of the soil led to the collapse of empires, from prehistory to the present.
Roulet, N., and Moore, T. M. 2006. Environmental chemistry: Browning the waters. *Nature* 444:283–284.
Recent carbon loss from British soils.
Stuerzenbaum, S. R., Andre, J., Kille, P., and Morgan, A. J. 2009. Earthworm genomes, genes, and proteins: The (re)discovery of Darwin's worms. *Proceedings of the Royal Society Biological Sciences, Series B* 276:789–797.
The taxonomy, ecology, distribution, and evolution of the world's worms.
Velando, A., Eiroa, J., and Domínguez, J. 2008. Brainless but not clueless: Earthworms boost their ejaculates when they detect fecund nonvirgin partners. *Proceedings of the Royal Society of London, Series B* 275:1067–1072.
Hermaphrodite worms assess mating history of partners before inserting sperm.
Wilkinson, K., Tyler, A., Davidson, D., and Grieve, I. 2006. Quantifying the threat to archaeological sites from the erosion of cultivated soil. *Antiquity* 309:658–670.
Patterns of loss of radioactive fallout reveal extent of disturbance by ploughs.

ENVOI: DARWIN'S ISLAND

Ashmole, P., and Ashmole, M. J. 2000. *St. Helena and Ascension Island: A natural history.* Oswestry, England: Anthony Nelson.
A tale of extinction and attempts at conservation on Darwin's penultimate island.
Causton, C. E., Peck, S. B., Sinclair, B. J., Roque-Albelo, L., Hodgson, C. J., and Landry, B. 2006. Alien insects: Threats and implications for conservation of Galapagos Islands. *Annals of the Entomological Society of America* 99:121–143.
One of the most important threats to Galapagos wildlife.
Ellison, A. M., Gotelli, N. J., Brewer, J. S., Knietel, J., Miller, T. E., Cochran-Stafira, L., Worley, A. C., and Zamora, R. 2003. Carnivorous plants as model ecological systems. *Advances in Ecological Research* 33:1–74.
Ecology and extinction risk of endangered insectivorous plants.
Ghazoul, J. 2005. Buzziness as usual? Questioning the global pollination crisis. *Trends in Ecology and Evolution* 20:367–373.
Opposed views on the fate of bees in the modern agricultural world.
Goulson, D. 2003. Effects of introduced bees on native ecosystems. *Annual Review of Ecology, Evolution, and Systematics* 34:1–26.
Positive and negative effects of the global spread of European bees.
Hendrix, P. F. 2006. Biological invasions belowground: Earthworms as invasive species. *Biological Invasions* 8:1201–1204.
Earthworms away from home as pests and agents of ecological change.
King, T. E., Parkin, E. J., Swinfield, G., Cruciani, F., Scozzari, R., Rosa, A., Lim, S. K., Xue, Y., Tyler-Smith, C., and Jobling, M. A. 2007. Africans in Yorkshire? The deepest-rooting clade of the Y phylogeny within an English genealogy. *European Journal of Human Genetics* 15:288–293.
Forgotten English outbreeding: a black Yorkshireman with many white descendants.

Kull, T., Kindlmann, P., Hutchings, M. J., and Primac, R. B. 2006. Conservation biology of orchids: Introduction to the special Issue. *Biological Conservation* 129:1–3.
The global plight of orchids and some possible solutions.

Lasker, G. W., and Mascie-Taylor, C. G. N. 2001. The genetic structure of English villages: Surname diversity changes between 1976 and 1997. *Annals of Human Biology* 28:546–553.
Increased outbreeding in the late twentieth century as shown by surname patterns.

Moore, L. T., McEvoy, B., Cape, E., Simms, K., and Bradley, D. G. 2006. A Y chromosome signature of hegemony in Gaelic Ireland. *American Journal of Human Genetics* 78:334–338.
Ancient differences in male mating success in Ireland leave traces today.

Prentis, P. J., Wilson, J. R. U., Dormontt, E. E., Richardson, D. M., and Lowe, A. J. 2008. Adaptive evolution in invasive species. *Trends in Plant Science* 13:288–294.
How invading plants such as the Oxford Ragwort evolve in their new homes.

Richardson, D. M., and Pysek, P. 2006. Plant invasions: Merging the concepts of species invasiveness and community invasibility. *Progress in Physical Geography* 30:409–431.
How some plants invade, and how some places can keep invaders out.

Scapoli, C., Mamolini, E., Carrieri, A., Rodriguez-Larralde, A., and Barrai, A. 2007. Surnames in western Europe: A comparison of the subcontinental populations through isonymy. *Theoretical Population Biology* 71:37–48.
High outbreeding in industrial countries measured by shared surnames on marriage.

Skirbekk, V. 2008. Fertility trends by social status. *Demographic Research* 18:145–180.
Reduction in variation in family size with increasing social equality.

Wade, M. J., and Shuster, S. M. 2004. Estimating the strength of sexual selection from Y-chromosome and mitochondrial DNA diversity. *Evolution* 58:1613–1616.
How the strength of sexual selection in humans has decreased over the centuries.

Index

aardvarks, 175
Abinger Hall, 180
acacia trees, 40–41, 42, 45
actinomycin, 43
Africa, 80, 27, 102, 103, 184; Ivanov expedition to, 24–25; life expectancy in, 201; and skin colour, 16, 17; soil depletion in, 184
Agent Orange, 122–23
agriculture: global crisis of, 197. *See also* farming
Akhenaton, 71
Alabama, 194
albinism, 16
alder trees, 43
alkaptonuria, 81
Alzheimer's disease, 12, 49
Amazon jungle, 117, 185, 195
amino acids, 16. *See also* DNA and genes
amusement, 53
amygdalae, 60–61, 66
Anatolia, 108
anatomy, 3; of barnacles, 134; of the ear, 144–45; Goethe's theory of, 141–42; skulls, 21, 143–44
Angola, 202
animals: below ground, 174–75; breeding of for food, 94–95; and convergent evolution, 27; dishonesty of, 162; extinction of, 191–92; as geological force, 167; inbreeding of, 76–77; as pollinators, 162; sexual habits of, 73, 85–88. *See also* endangered species; *and names of specific animals*
annelids, 169. *See also* worms
anteaters, 175
antibiotics, 43, 174
ants, 40–42
apes, 14, 23; common ancestry with humans, 1–2, 9–10; emotions of, 53, 54; facial expressions of, 53; impact of habitat destruction on, 198. *See also* gorillas; orang-utans; primates
apples, 98
Arabian oryx, 76–77
Arabidopsis, 5
archaea, 174–75
archaeology, 166, 168–69, 178–80
Arctic, 155
Ardipithecus, 9
Argentina, 195
Aristotle, 152, 172
armadillos, 175
arthropods, 131–32. *See also* barnacles
artificial selection, 95
Ashdown Forest, 190
Asia, 16–17. *See also* China; India; Japan
Asperger's syndrome, 64
Australopithecus afarensis, 9
autism, 63–67
Autobiography (Darwin), xiii, 167
auxins, 121, 122

bacteria, 176; and antibiotics, 174; behaviour of, 49; interaction of with plants, 40, 42–45

badgers, 175
Baer, Karl von, 131
Bagehot, Walter, 81
Bakewell, Robert, 95
barley, 92, 106
barnacle geese, 130–31
barnacles (cirripedes), xii, 131–41; anatomy of, 134; classification of, 137–38; diversity of, 131–32, 137–38, 139; as environmental threat, 195; evolution of, 137–39; fossils of, 137–39; larvae of, 137, 139–40, 147; as model organism, 132; as parasites, 136–37; resemblance of to other creatures, 149; sex life of, 135
Barnacles (Darwin), 133–34
Bartram, William, 31
basset hounds, 52
Bateson, William, 142
bats, 152
Beagle, HMS, ix, x, xiii, 96, 132–33, 136, 180, 188–89
beans, 162
Beaulieu Abby, 179
beer, 107–8, 114
bees, 74; environmental threats to, 196–97; as pollinators, 152, 157, 158, 160–61
beetles, 157
Bell, Charles, 49
Belt, Thomas, 40–41
Belyaev, Dmitry, 105
Bentham, Jeremy, 4
Bettelheim, Bruno, 66
bindweed, 118
bin Laden, Osama, 202–3
Biographical Sketch of an Infant (Darwin), 62, 65
biomantle, 174
biosphere, 192. *See also* environmental devastation
birds, 162; evolution of, 146; migratory, 125–26. *See also* barnacle geese; chickens; finches; sparrows
bitter vetch, 92
blackberry, 194
Black Death, 18
black walnut, 127
bladderworts, 31–32
bloodhounds, 87
blue plumbago, 39
blushing, 54–55, 65
bonobos, 12, 53
Border collies, 51
Bose, Chandra, 113
botany. *See* insectivorous plants; orchids; plants
brains: of humans, 20, 21–22, 46–48, 60; physiology of, 48–49. *See also* emotions
Brazil, 175, 185
Britain: archaeological finds in, 166, 168–69, 178–80; Darwin's travels in, xiii–xiv; ethnic minorities in, 200–201; intermarriage in, 200–201; invasion of pests in, 193; life expectancy in, 201; marriage laws in, 71; as microcosm of global upheaval, 189–91; obesity in, 93; population growth in, 187; sexual inequality in, 203
British Association Barrow, 179–80
bromeliads, 32–33
Brooke, William, xiv
bulldogs, 100
bull terriers, 51, 52
Buffon, Comte de, 173
Bush, George W., 56
butterwort, 30, 37

cactuses, 28
calcium, 93
Cambrian era, 139
Canada, 185, 195
cancer, 94
canines. *See* dogs; foxes; wolves
carbon, 176, 182, 194–95
carnivorous plants. *See* insectivorous plants
Carolinas, 31, 195
cats, 191–92
cattle. *See* cows
cereal diet, 106–7
cesium, radioactive, 182
Chaco Canyon, 183
Chesterfield, Lord, 116
chickens, 5, 103
chickpeas, 92, 99
Chihuahuas, 101
children: facial expressions as understood by, 57; feral, 63; obesity in, 93–94
Chile, 117, 132–33, 139, 195

chimpanzees, 53, 65; as compared with humans, 12–13, 22–23, 24; diet of, 13; hybridising with humans proposed, 25
China, 16, 34, 183, 184–85
Chomsky, Noam, 23
chromosomes, human, 7, 13, 20. *See also* DNA and genes
clematis, 116
Cleopatra, 172
Climbing Plants. See *Movements and Habits of Climbing Plants, The*
cobra lily, 32, 33, 35
Cocker spaniels, 52
cod-liver oil, 18
common ancestry, 28; of apes and humans, 1–2, 9–11, 14–15; of barnacles, 133; and Neanderthals, 10–11
Common Plants Survey, 190
conservation movement, 155, 184
convergent evolution, 27–28, 45
convolvuli, 117
corn. *See* maize (corn)
Cornovii, 178
corn syrup, 94
cougars, 76
cows: domestication of, 102–3; worship of, 102
cowslips, 83–84
crabs: and barnacles, 136, 137, 139
Crick, Francis, xiii
crops, domestic, 95–99
Cross and Self-Fertilisation (Darwin), xi, xii, 71
crustacea, 138, 147. *See also* barnacles; crabs; lobsters
cryptochromes, 124, 126
cuckoos, 190
Culpeper, Nicholas, 150
cystic fibrosis, 81

dachshunds, 101
dance flies, 162–63
Darley, Thomas, 103
Darwin, Annie (daughter), 70, 83
Darwin, Charles: health problems of, xiv, 115; misconceptions about, ix–x; travels of, xiii–xiv; works by, xi–xii. *See also names of individual titles of works*
Darwin, Charles (son), 69–70
Darwin, Emma Wedgwood (wife), 28, 69, 168
Darwin, Erasmus (brother), xiv
Darwin, Erasmus (grandfather), 72
Darwin, Francis (son), 50, 89, 115, 126
Darwin, George (son), 70, 77–78, 79, 89
Darwin, Henrietta (daughter), 70
Darwin, Horace (son), 89, 171, 178
Darwin, Leonard (son), 70, 89
Darwin, Robert (father), 69
Darwin, Susannah Wedgwood (mother), 69
Darwin, William (son), 89, 62
dates, 152
deafness, 78
Dee Estuary (Wales), 181–82
Deere, John, 184
deficiency diseases, 18, 93, 97
depression, 62
Descent of Man, The (Darwin), xi, xii, xiii, 2, 3, 15, 23, 25, 150, 151
diabetes, 94, 110–11
Dickens, Charles, 67
diet: deficiencies in, 93; of *Homo sapiens*, 92–93. *See also* farming; food; obesity; *and names of specific foods*
Different Forms of Flowers on Plants of the Same Species, The (Darwin), 83–84
digestive processes: of insectivorous plants, 36–38, 39; of worms, 170
dinosaurs, 26, 34, 159
dioxin, 122
directed evolution, 91
discoverists, 166, 168
disease: and dietary deficiencies, 93; and natural selection, 18–19, 89
Disraeli, Benjamin, 1
dissection, 3
DNA and genes, xiv, 131; and autism, 66–67; of cattle, 102–3; of chimpanzees, 12–13, 198; of corn, 97–98; of cows, 102–3; of dogs, 52, 100; and evolutionary theory, 4–5, 32; and farming, 95; and fossils, 137, 159; and Fragile X syndrome, 67; in hermaphrodites, 85; human, 4–8, 11, 13, 15–16, 108; and the immune system, 88–89; of maize, 97–98; and migra-

DNA and genes (continued)
tory patterns, 199–200; and molecular genetics, 137; and obesity, 109–10; and paternity, 163–64; of pigs, 103; similarity of across species, 14, 131, 143, 148–49; and skin colour, 15–17; structure of, 5, 6–8; and surnames, 78–79; touch genes, 128; unity of type in, 131, 149. *See also* mutations, genetic; natural selection; variation and diversity
Dobbs, Arthur, 31
Dobermans, 52
dodders, 126
dogs, 50–51; domestication of, 99–102, 104–5; personalities of, 51–52
dolphins, 136
Dominican Republic, 159
Down House, x, 154, 169, 173–74, 193
Doyle, Arthur Conan, 26, 30
Duchenne de Boulogne, Guillaume-Benjamin-Amand, 58, 59
dust bowl, 184
dyspepsia, xiv

ear: anatomy of, 144–45, 146–47; evolution of, 146–47
earthworms. *See* worms
earwig, 189
Ebola virus, 198
Ecuador, 138
Egypt, 102, 172
electricity, 35, 37, 119
electro-encephalogram (EEG), 91–92
elephantiasis, 169
elephants, 27, 103, 153
embryology, 130–31, 140–41, 142–43; and ears, 146–47; and fossils, 144
emotions: of humans, 46–48, 59–60; of humans as compared with other species, 48, 49–50, 52–53, 54; medical technology applied to interpretation of, 58–60
endangered species, 76–77, 188–89, 191–93, 196–97
England. *See* Britain
environmental devastation: caused by humans, 191–92, 198–201; and soil protection, 185; in the twentieth and twenty-first centuries, 188–92
epidemics: impact of on human populations, 18–19
epilepsy, 64
equilibrium, 124
erosion, 173
ethylene, 126
etiquette, 54
eugenics, 69, 70
euphorbias, 28
Europe: farming in, 11, 106; inbreeding among aristocracy in, 79; marriage patterns in, 80; and skin colour, 16, 17
evolution, theory of, 26, 131; and embryology, 131; evidence for, 2–3, 141, 146; and the fossil record, 141, 146; and genetic trends, 198–99; and *Homo sapiens*, 2, 25, 198; and molecular biology, 3; and natural selection, 2, 15, 27, 146; opponents of, 49; and sexual conflict, xii, 151–52, 163–65; technology as applied to, 3–5, 8. *See also* convergent evolution; DNA and genes; natural selection; variation and diversity
evolutionary developmental biology, 133
Expression of the Emotions (Darwin), xi, xii, 47–48, 52, 54, 58, 59–60

face-blindness, 57
facial expressions, 47–49, 54–57; and autism, 65, 66; and children, 57; cultural differences in interpretation of, 56–57; of dogs, 50–51; interpretation of, 54–58
farming: and animals, 99; in Europe, 11, 106; history of, 11, 91–92; and human population explosion, 18; in the Middle East, 92, 106, 183; nitrogen as essential to, 42–43, 44; pure lines as used in, 75; scientific, 91, 95, 97; and soil depletion, 181–85; transformation of, 90–93
Farrer, Thomas, 178–79
ferns, 116
fertilisation: process of, 142. *See also* pollinators; sex
fertiliser, 44–45, 176
fertility, 111, 203–4
finches, ix, 76
fireflies, 163
fish, 16, 89, 189; fossils of, 145; gills of, 145–46
Fitzroy, Robert, x

flies, 28, 153, 162–63. *See also* insectivorous plants
flight: evolution of, 27
Florida, 158, 194, 195
flowers: scarcity of, 190; fossils of, 157, 159; and insects, 151, 153. *See also* orchids; plants
flytraps. *See* insectivorous plants; Venus flytraps
foliar feeding, 39
food, 90–99; cereals, 106–7; cost of, 91; cultivation of, 91–92, 94–99; meat consumption and production, 94–95, 99, 102–3; and natural selection, 107–8. *See also* diet; junk food; obesity; *and names of specific foods*
Formation of Vegetable Mould by Earthworms (Darwin), xi, xii, 166–68, 171, 176. *See also* soil; worms
fossils, 141; of barnacles, 137–39; of early vertebrates, 145–46; of flowers, 34, 157, 159; of insectivorous plants, 34; of primates, 8–11, 14; soil, 173; of water lilies, 157
foxes, 105–6
fragile X syndrome, 67
France, 81
Franklin, Benjamin, 125
functional magnetic resonance imaging (fMRI), 58–59
fungi, 33, 43
fungus gnats, 161

Galapagos Islands, 191–92
Galen, 4
Galton, Francis, 47, 69, 74, 177
Garrod, Sir Archibald, 81
genome. *See* DNA and genes; human genome project
Georgia (USA), 194, 195
German shepherds, 52
Gilbert, W. S., 1
giraffes, 41, 152
Gladstone, William Ewart, 55
global warming, 176
goats, 191
Goethe, Johann Wolfgang von, 141–42
Gondwana, 14
gophers, 175
gorillas, 12, 46, 53, 54, 198
grains, 106, 107. *See also* barley; wheat
Grant, Robert, 132
Gray, Asa, 193
greyhounds, 51
Guinea, 25
Guy the Gorilla, 46

hair color, 106, 107
Harvey, William, 6
heart, 4
heart disease, 82, 94
Heidelberg Man, 10
hemoglobin, 19
hereditary disorders, 81–82. *See also* inbreeding
hermaphrodites: animals, 73; plants, 72–75, 76, 85
Herodotus, 172
Hindus, 79–80
Hippocrates, 94
Holt, Vincent, 38
homeobox genes, 147–49
Homo antecessor, 10
Homo erectus, 10
Homo habilis, 9–10
Homo sapiens, xii; brains of, 20, 21–22, 46–48, 60; as compared with apes, 1–2, 11–14, 12, 13, 22, 23, 54, 108–9; diet of, 92–93; embryo of, 140–41; as endangered species, 198, 201–2; facial expressions of, 47–48, 54–57; future of, 188, 198–201, 204; homogenization of, 191, 199–201; impact of epidemics on, 18–19; learning ability of, 22; migratory patterns of, 11, 200–201; moral sense of, 2, 25; and natural selection, 201–4; scent as signal to, 87–88; and sexual deception, 164–65; skin colour of, 15–18; social nature of, 22, 46–47; uniqueness of, 25; vertebrae of, 141–42, 143. *See also* emotions
honey, 152
honeybees, 157, 196
honeysuckle, 118
Hooker, Joseph, 193
hops, 107–8, 114–15, 116, 119, 125, 190
hormones, xii; in plants, 115, 121–22; and sexuality, 153, 163; testosterone, 153

horse breeding, 95, 103
human beings. See *Homo sapiens*
human genome project, 4–5
Huxley, Thomas Henry, xi, xii, xiv, 130
hybridisation, 24–25
hypothalamus, 104

Iceland, 82
inbreeding: of animals, 76–77; Darwin's concerns about, xi, 69–70, 77, 89; health problems stemming from, 80–83; of humans, 69–71, 77–81, 89, 199; of plants, 71–76
India, 79–80, 83, 176, 204
infants, 62, 65
infrared radiation, 123
insectivorous plants, 26–27, 28–45; bacteria as used by, 40, 42; damage to habitat of, 192; Darwin's experiments with, 36–37, 38–39; digestive processes of, 36–38; early fossils of, 34; and natural selection, 28, 29, 30, 39–40, 41, 45
Insectivorous Plants (Darwin), 28, 30
insects, 138; dishonesty of, 162–63; and flowers, 151–52; as pollinators of orchids, 152, 154–58. *See also* insectivorous plants
Iran, 80
Irish wolfhound, 101
iron deficiency, 93
Ishtar, 102
Isle of Wight, 179
Israelites, 102
Itard, Jean Marc Gaspard, 63–64
Ivanov, Il'ya Ivanovich, 25
ivy, 116, 194

jackdaws, 190
Japan, 16, 83, 194, 195, 201
Jefferson, Thomas, 31
Johnson, Samuel, 72
junk food, 93, 96

Kazakhstan, 98, 103, 181
Keith, Sir Arthur, 178
Kent, England: changes in since Darwin's time, 187–88, 189–90
King Charles spaniels, 101
Kingsley, Charles, 12
knapweed, 194
Knome Corporation, 4
knotweeds, 39
kudzu, 194

lancelets, 148
language and speech: origin of, 23–24
Lascaux cave-paintings, 102
laughter, 53
Le Brun, Charles, 48
lemurs, 14
lentils, 92, 99
Leonardo da Vinci, 167
Leviticus, 71
lianas, 116–17, 195
Liberia, 202
lice, 20–21
life expectancy, human, 201–2
Linnaeus, Carl, 31, 32, 63, 120
lizards, 146
loblolly pines, 75
lobsters, 132, 139, 147
Lombroso, César, 31
Lubbock, John, 179
Lucy, 9
Lyell, Charles, 183
Lyte, Henry, 30

macaques, 12, 14, 53, 54
Mackay, Charles, 47
magnetism, 125–26
maize (corn), 92, 93; cultivation of, 94, 96–97
malaria, 18–19
manatees, 175
mapping, xiii
marmosets, 87
marmots, 175
marriage: between cousins, 77–78, 80, 83; changing patterns in, 199–201; diversity of customs relating to, 79–80; laws affecting, 71, 77; social pressures affecting, 86
Marx, Karl, 182
mass hysteria, 47
Maya, 138
meat production, 94–95, 99, 102–3
Medawar, Peter, xiii

medical technology, 58–59
meerkats, 27, 175
melanism, 16, 17, 27. *See also* skin colour
Mesa Verde, 183
Mesmer, Franz Anton, 125
Mesopotamia, 183, 185
metal detectors, 166, 168
Mexican hairless dogs, 101
mice, 24, 86–87, 89, 124
Middle East, 80; cows in, 102; farming in, 92, 106
migration, human, 11
milk, 95, 108
mimosas, 114, 119–20, 128
Minnesota, 196
Minotaur, 102
Miocene era, 9
Mississippi, 194
Mitchell, Sir Arthur, 70
mites, 73
model organisms, 132
molecular biology, 3; as applied to agriculture, 95
molecular genetics, 137–38
moles, 175
monkey cups, 32
monkeys, 12, 14, 53, 60–61
Moray, Sir Robert, 130, 134, 149
Mormons, 82, 196
morning glory, 74, 75
moths, 162
Moussaoui, Zacarias, 46, 68
Movement in Plants. See *Power of Movement in Plants, The*
Movements and Habits of Climbing Plants, The (Darwin), 114
muesli, 107
mushrooms, 33
music, 126, 147
mustard plant, 118
musth, 153
mutations, genetic, 16, 18; and natural selection, 19–20, 27; in plants, 118

narcolepsy, 52
narwhals, 175
National Trust, 190
Native Americans, 183; diet of, 94
natural selection, 2–3, 6, 90, 173; and barnacles, 140; and climbing plants, 115; by diet, 107–8; and disease, 18–19, 89; evidence for, 27; and genetic mutation, 19–20, 27; and genetic variation, 15, 18, 107–8; and *Homo sapiens*, 201–4; and insectivorous plants, 28, 29, 30, 39–40, 41, 45; and obesity, 111; and orchids, 158–59; and sex, 83, 111, 150; and sexual dishonesty, 154, 162–65; and skin colour, 15–18; and vertebrates, 146
Nauru, 111
Neanderthals, 10–11
nematodes, 33
Nepenthe, 32
nerve transmission. *See* neurotransmitters
neuroscience, 58–60
neurotransmitters, 61–62, 66, 104–5
New Zealand, 181, 193
Niger, 185
Nile River, 172
nitrogen: as absorbed by plants, 29, 33–35, 39–40, 42–43; and bacteria, 29; as used in farming, 42–45
nitrous oxide, 176

obesity: in children, 93–94; environment as factor in, 110–11; fertility as affected by, 111; genetics as factor in, 109–10; growing prevalence of, 90–91, 93–94
obsessive-compulsive disorder, 52
O'Donnell, Lord Turlough, 203
Omer Ibn Al-Khatab, 81
orang-utans, 1, 3, 46, 53, 62
orchids: deceptive nature of, 153–54, 160–62; habitat of, 155–56; pollination of, 152, 154–58; seeds of, 156; and sexual signals, 150, 153; threats to, 192–93
Orchids and Insects (Darwin), 70, 151
Orgasmusgesicht ("orgasm countenance"), 53
Origin of Species, The (Darwin), ix, x–xi, xii, 4, 8, 70, 131; and barnacles, 133
Overton Down, 180
Oxford ragwort, 194
oysters, 189

Pakistan, 80, 83
panthers, 76

Papua New Guinea, 155
parasites, 161, 169; barnacles as, 136–37
parthenogenesis, 72–73
Peacock, Thomas Love, 23
peacocks, 155
peas, 92, 99, 162
pellagra, 97
PET scans, 58–59
phenylketonuria, 14
pheromones, 157
phototropins, 123
phrenology, 49–50
phytochromes, 123–24
pigs: domestication of, 103; as predators, 191
Piltdown Man scandal, 178
Pima Indians, 94
pines, 127
pit bulls, 51
pitcher plants, 32–34, 35–36, 38
plague, 18
Plague of Justinian, 18
plants: and ants, 40–42; bacteria as used by, 40, 42; biological clock in, 120; botanical intelligence of, 113–15, 119; breeding of for food, 95; climbing, 113, 114–15, 116–19, 194–95; environmental threats to, 195–96; and foliar feeding, 39; hormones in, xii, 115, 121–22; inbreeding of, 71–72, 83–84; and light, 116, 117–18, 120–21, 123; movement in, 113–15; and natural selection, 115; roots of, 115, 116, 118, 124–25, 127, 174, 175; self-fertilisation of, 71–75; and sex, 71–75, 83–85, 150–59; sexual dishonesty of, 150, 153–54, 160–62; signals sent by, 127. *See also* flowers; insectivorous plants; orchids; pollinators
Plato, 183
plow, 166; and soil depletion, 181–82, 183–84
pointers, 51
poison oak, 118
poisons, 122, 127, 136, 194; and insectivorous plants, 30, 36–37
pollen, 85
pollinators: dishonesty of, 160–65; environmental threats to, 196–97; history of, 159–60; insects as, 38–39, 152, 154–58; and orchids, 152, 154–58; and plants, 151–52, 158
population growth, 187, 188, 203–4
Portable Antiquities Scheme, 169
positron emission tomography (PET), 58–59
potatoes, 99
Power of Movement in Plants, The (Darwin), xii, 114, 115, 119
prairie dogs, 175
predators: as threat to the environment, 191–92
primates: emotions as displayed by, 53, 54; fossil record of, 8–11, 14. *See also* apes; chimpanzees; *Homo sapiens*; monkeys
primrose, 84
primulas, 39
proteins, 16, 118, 122, 123; in barnacles, 135; and DNA, 19–20; and scent, 87
psychology, 48. *See also* brains; emotions
pugs, 100
pure lines, 75
Pyrenean mountain dogs, 51

Quantock Hills, 182

rabbits, 14
rainbow plants, 30
rats, 87
Red Crag deposits, 138
Red List of Threatened Species, 197
religion, ix, x, xii–xiii
retrievers, 51
rhesus macaques, 12, 14
rice, 92, 99
rickets, 17–18
Roman Empire, 178–79, 182–83
Romanes, George, 48
Roosevelt, Franklin D., 184
Roraima, Mount (Venezuela), 26, 27, 29, 32
roses, 116
rottweilers, 52
Rousseau, Jean-Jacques, 47, 63
Royal Society, 132
Royal Society for the Protection of Birds, 190
Ruskin, John, 28, 151
Russia, 24–25, 105, 173
rye, 106

sagebrush, 127
Sahel (Africa), 184

Saint Helena, 188–89
Sanderson, Burdon, 37
Sartre, Jean-Paul, 47
Saudi Arabia, 81–82
scent: and mating habits of animals and humans, 86–87; and pollination of plants, 126, 157, 158, 160–61
Scotland, xiii, 70, 107, 127
seasonal affective disorder, 124
Sedgwick, Adam, xiii
segmented creatures, 140–41, 148
self-fertilisation, 70; of plants, 71–75, 84–85. *See also* inbreeding
serotonin, 61–62, 66, 104–5
setters, 51
sex: and animals, 73, 85–88; and barnacles, 135; and chimpanzees, 12; and evolution, 83; within families, 70–71, 83; legal prohibitions affecting, 71, 77; and natural selection, 83, 111, 150, 202–3; and plants, 71–75, 83–85, 150–59; scent as factor in, 87–88; and worms, 171. *See also* inbreeding; marriage
sexual signals: dishonesty in, 150, 152–53, 160–65
Shakespeare, William, 48–49, 60
sheep, 76, 95, 104
Shelley, Percy Bysshe, 114, 115, 128–29
shrimp, 132, 175
sickle-cell disease, 19, 20, 81
signalling theory, 160
silver foxes, 105–6
Simpson, O. J., 18
skates, 175
skin colour, 15–18
skulls, 21, 143–44
slugs, 74
smell. *See* scent
Smiles, Samuel, xi
snails, 73
Soay sheep, 76
sodium, 93
soil: beneath the sea, 175–76; biodiversity in, 174–75; depletion of, 181–85; protection of, 185; science of, xii, 167–68, 174–75, 185–86; and wind, 180–81; and worms, 166, 167, 170, 172–75, 176–78, 185
solitary confinement, 67–68
somites, 142–43
song sparrows, 76
Spain, 138
sparrows, 76, 190
Spencer, Herbert, x, xiv
spiders, 160
Spitz dogs, 51
springer spaniels, 52, 104
Staffordshire gold, 166, 169
Stonehenge, 168
strangler fig, 118
streptomycin, 174
stroke, 94
sugarcane, 127
sugars, 174
Sumerians, 102
sundews, 28, 30, 33, 34–35, 36–37, 38, 39, 192
sunflowers, 120
surnames, 7, 78–79, 199–200, 203
"survival of the fittest," ix–x
Swaziland, 202
Sweden, 77, 163–64
symbiosis: of ants and plants, 40–42
Syria, 92

tamarins, 87
tamarisks, 39
taxonomy, 137
Tennyson, Alfred, x
teosinte, 96, 97
terra preta, 185
testosterone, 153
"Thousand Genomes Project," 7
toadflax, 73
tobacco, 127
tortoises, 191
touch, sense of: in plants, 115, 128
touch genes, 128
trees, environmental threats to, 196
tropical forests, 116–17
tuberculosis, 174
Twain, Mark, 54–55, 184
twins, 109–10

United States: dust bowl, 184; earth mounds in, 175; food and diet in, 93–94; homogeneity in, 200; invasion of pests in, 193;

United States (continued)
kudzu in, 194; life expectancy in, 201; marriage laws in, 77; obesity in, 91, 93–94
Universal Soil Conservation Law, 184
University College London, 115, 132
Ur of the Chaldees, 183
Uzbekistan, 181

variation and diversity, 8, 131, 148, 198; in barnacles, 131–32, 136, 137; in dogs, 51–52, 99–101; in insectivorous plants, 30, 33; in plants, 116, 158–59; repeated structures as aspect of, 142–43, 145–46, 149
Variation of Animals and Plants Under Domestication, The, 90, 99, 102
vegetable mould. See *Formation of Vegetable Mould by Earthworms* (Darwin); soil; worms
Venezuela, 26, 170
Venter, Craig, 6–7
Venus flytraps, 31, 33, 35, 39, 192
vertebrae: human, 141–43
vertebrates, 143, 145–46
Vesalius, 4
Victoria, Queen, 1
Vietnam War, 122–23
vines, 113
Virgil, 152
virgin birth. *See* parthenogenesis
Virginia creepers, 116
Virgin Mary's socks, 40
Viroconium, 178
vitamin D, 17–18, 107
voles, prairie, 163–64

Wales, xiii, xiv, 169
Wallace, Alfred Russel, xiv, 170, 175
wasps, 153, 156
water buffalo, 103
water lilies, 157
waterwheel plants, 31
Watson, James, xiii, 4, 5
waves, detection of, 145–46
Weddas of Ceylon, 55
Wedgwood, Josiah, 69, 168
Wedgwood, Sarah (Darwin's grandmother), 69
Wedgwood, Sarah Elizabeth (Darwin's sister-in-law), 28, 190
weeds, 185, 193–94
Wellcome, Henry, xiv
Westminster Abbey, x, 186
whales, 146; and barnacles, 136
wheat, 92, 99, 106
whippets, 101
whistling thorn, 41
White, Gilbert, 172
Wilde, Oscar, 70
Wilde, Sir William, 70, 78
wild peas, 162
Wilson, Harold, 168
wind, 180–81
witchweeds, 127
Wolsey, Cardinal, 49
wolves, 99, 100, 104
wombats, 175
Woodland Trust, 190
worms, xi, xii, 73, 166–67, 178–79; anatomy of, 166, 169–70, 171; and Darwin's experimental millstone, 177–78, 179; diet of, 172; digestive process of, 170; as environmental threat, 195–96; excretions (casts) of, 170, 172, 176–77, 178, 179, 180; intelligence of, 171–72; and regeneration, 170–71; sex life of, 171; and soil, 166, 167, 170, 172–75, 176–78, 185; species of, 169
Wyman, Bill, 168

Xoloitzcuintli, 101

yawning, 65–66
Yellow River (China), 183
Yixian fossil bed, 34
yuccas, 162

zebra fish, 16
zoos, 1, 46, 53, 103